T3-BSK-775

A Healthy Old Age
A Sourcebook for Health Promotion with Older Adults

Revised Edition

The *Journal of Gerontological Social Work* series:

- *Gerontological Social Work Practice in Long-Term Care,* edited by George S. Getzel and M. Joanna Mellor
- *A Healthy Old Age: A Sourcebook for Health Promotion with Older Adults (Revised Edition),* by Stephanie FallCreek and Molly Mettler
- *The Uses of Reminiscence: New Ways of Working with Older Adults,* edited by Marc Kaminsky
- *Gerontological Social Work Practice with the Community Elderly,* edited by George S. Getzel and M. Joanna Mellor

A Healthy Old Age
A Sourcebook for Health Promotion with Older Adults

Revised Edition

Stephanie FallCreek and Molly Mettler

The Haworth Press
New York

A Healthy Old Age: A Sourcebook for Health Promotion with Older Adults, Revised Edition has also been published as the *Journal of Gerontological Social Work*, Volume 6, Numbers 2/3, November 1983.

The first edition was published in 1982 by the Health Promotion with the Elderly Project, Center for Social Welfare Research, School of Social Work, University of Washington, Seattle, Washington. Its preparation was made possible in part by a grant from the Department of Health and Human Services Administration on Aging (Grant #90-AT-0016).

The Haworth Press, Inc., 28 East 22 Street, New York, NY 10010

Library of Congress Cataloging in Publication Data

FallCreek, Stephanie.
A healthy old age.

"Has also been published as the Journal of Gerontological social work, volume 6, numbers 2/3, November 1983."—Verso t.p.
Includes bibliographical references and index.
1. Community health services for the aged. 2. Aged—Care and hygiene. 3. Health education. 4. Preventive health services. I. Mettler, Molly. II. Title. [DNLM:
1. Geriatrics—Handbooks. 2. Health promotion—In old age—Handbooks. 3. Health services for the aged—Handbooks.
WT 100 F194h]
RA564.8.F34 1984 613'.0438 83-18414
ISBN 0-86656-247-8

A Healthy Old Age

A Sourcebook for Health Promotion with Older Adults

Revised Edition

Journal of Gerontological Social Work
Volume 6, Numbers 2/3

CONTENTS

RESOURCES

RENEE SOLOMON, DSW, *Associate Professor, Columbia University School of Social Work, NY, NY*
SHELDON TOBIN, PhD, *Director, Ringel Institute of Gerontology, Albany, NY*
TERESA JORDAN TUZIL, MSW, *Consultant on Aging, New York, NY*
EDNA WASSER, MSW, *Consultant, Fellow, Gerontological Society, Miami, FL*
MARY WYLIE, PhD, *Professor, Department of Social Work, University of Wisconsin, Madison, WI*

FROM THE EDITOR

We are pleased to be the publisher of *A Healthy Old Age: A Sourcebook for Health Promotion with Older Adults.* We think the time is ripe for a publication of this *genre*, and we think also that the authors and their associates have done an excellent piece of work.

Health promotion activities are becoming important elements in the programs of many agencies in the aging network, and for good reason. The evidence is persuasive that all of us—not just older people—can improve our chances of reasonably healthy old ages if we change or improve some of our habits. As the authors point out, exercise, proper nutrition, environment awareness, no smoking, stress management, etc., *can* make a difference, and the program described in these pages is an exciting approach to the how to of life style changes.

Health promotion programs, in our judgement, belong to no one profession; rather they are enterprises which require the work of a team approach, with social work being one member of the team.

We commend this publication to you.

Rose Dobrof
Editor

Preface to the Revised Edition

One year, more than twenty workshops and presentations later, we find *A Healthy Old Age* is being used in a variety of settings across the nation, from nursing homes to the *YWCA*, in public health departments, and senior centers, by Red Cross Chapters and private hospitals. Each agency finds it useful for different reasons and the health promotion programs which rely upon it for information and assistance are truly varied. We are pleased and grateful by the wide acceptance the Sourcebook has found.

We have also learned a lot since we developed the first edition of the Sourcebook and this edition reflects some of that learning. A few more programs around the country are identified, as are some additional resource materials. Some sections of the sourcebook needed clarification and we have tried to accomplish that. The nutrition section, particularly, has been the subject of some controversy. The Gerontological Nutrition Dietetic Practice Group of the American Dietetic Association has expressed concern with some of the terminology and a few of the references used in the nutrition curriculum modules. We appreciate these concerns and have addressed at least some of them in this section.

An additional three hundred pages would be required to incorporate the ideas, materials and experiences which have been shared with us by those of you implementing health promotion programs with older people since the publication of the first edition. Thank you all for your comments, suggestions, enthusiasm and inspiration. The more we share our experiences and resources the better able we will be to develop and deliver the kinds of program that result in increased independence, strength and joy for all who participate.

Stephanie FallCreek
Las Cruces, NM

Molly Mettler
Devon, England

Preface to the First Edition

The Health Promotion with the Elderly Project is a two-year continuing education and training grant funded by the Administration on Aging and administered by the University of Washington School of Social Work in conjunction with the Institute on Aging, Seattle, Washington. The purpose of the project is to develop and distribute information about health promotion for older adults through educational materials and regional training workshops.

The project has developed *A Healthy Old Age: A Sourcebook for Health Promotion with Older Adults* for trainers, program planners and service providers who are interested in developing or expanding health promotion programs with the elderly.

Many dedicated people gave freely of their time and talents in working with us to create *A Healthy Old Age*. We would like to acknowledge the institutional support of the School of Social Work, Scott Briar, Dean. A special thank you to Michael J. Austin, Director of The Center for Social Welfare Research; Alice Kethley, Deputy Director of The Institute on Aging, and HPE Co-Project Director; and James Steen, our Project Officer at the Administration on Aging for their guidance and support.

The development of the sourcebook owes a great deal to the collaboration and enthusiasm of the Wallingford Wellness Project staff: James Barrett-DeLong; Lisa Christopherson; Bernadette Lalonde; Kelley Reid; Sue Stam; Jackie Williams and to the Wellness Project participants (ages 13-84) who "fine-tuned" much of the information and activities contained in this manual.

For their help in compiling resources and shaping ideas, we wish to thank Marilynne Gardner, Helen Hardacre, Kari Knutson, Marsha McMurray-Avilar, Cindy Moe, Gloria Montelongo, Jody Skinner and Raymond Unks. Special thanks to Betty Blythe for her constructive criticism and editorial assistance and to Dorothea Hayes and Gene Hooyman for their assistance with the rewrite. We are also grateful for the fine artwork, supplied by Tommy Adams.

We would like to thank the many dedicated professionals from both the aging and health services networks who volunteered their time, resources and expertise to the project. We would also like to acknowledge the input and gentle hints of the training workshop participants in Seattle and Boston who worked with us on the draft copy of the Sourcebook.

Anna Bolstad, Joan Hiltner, Suzanne Mackler and Cheryl Yates offered valuable assistance throughout the project with their skills in word processing. Jim Goll of the University of Washington Department of Printing led us gracefully through the publishing maze.

And to HPE core staff, Mary Alderete and Dorothea Hayes, warm gratitude is extended for their contributions to the Sourcebook *and* their efforts in keeping the project afloat and on track.

SPECIAL MENTION

We would like to express our appreciation to the hard-working, deeply caring and often-challenging members of the Health Promotion with the Elderly Project's two advisory committees: the National Review Panel and the Local Advisory Council. Their collective vision and commitment to excellence continues to be invaluable.

National Review Panel

Donald Ardell
Planning for Wellness

William Beery
University of North Carolina

Ruth Behrens
American Hospital Association

Edwin Bierman, M.D.
University of Washington

Rick Carlson
Health Resources & Comm., Inc.

Pearl German
Johns Hopkins University

Lawrence Green
University of Texas

Nathan Maccoby
Stanford University

David Maldonado
University of Texas

William McMorran
NRTA/AARP

Nancy Milio
University of North Carolina

Terry Monroe
Wellness Resources, Inc.

Henry Peters
University of Alabama

Ann Radd
National Assoc. of Human Dev.

Maureen Henderson, M.D., Chair
University of Washington

Joyce Leanse
National Council on the Aging

Barry Lebowitz
National Institute on Mental Hlth.

Lowell Levin
Yale University

Terry Lierman
Carley Capital Group

John Lowe
Temple University

Kelley Reid
Wellness Resources Group

James Rua
New York State Off. for the Aging

Keith Sehnert, M.D.
Vinland National Center

James Steen, Ex Officio
Administration on Aging

Anne Zimmer, Ex Officio
Administration on Aging

Local Advisory Council

Michael Austin, Chair
University of Washington

Leo DesClos
Senior Services and Centers

Miriam Gray
Institute on Aging

Gene Hooyman
Providence Hospital

Denise Klein
Division on Aging

Betty Mathews
University of Washington

Kay Osborne
Nutrition Consultant

Jeremy Sappington
Group Health Cooperative

Sue Stam
Wellness Resources Group

Stephanie FallCreek
Molly Mettler

How to Use the Sourcebook

The purpose of *A Healthy Old Age: A Sourcebook for Health Promotion with Older Adults* is to provide information about health promotion program planning, activities, and resources to people who are interested in developing or expanding wellness programs with older adults.

We hope that *A Healthy Old Age* will be used somewhat like a cookbook for preparing health promotion programs for older people. Those of you who are in the beginning stages can turn to the sourcebook for the basics: a description and a rationale for health promotion programming with your population; program planning suggestions; key ingredients for program content; sources of support for material and human resources; and potential problems, challenges, and barriers to anticipate and overcome.

For the gourmet chefs among you, the sourcebook can be used as a ready reference guide, a source of suggested menus and a checklist of basic ingredients for particular recipes. You may already be familiar with delivering nutrition and physical fitness programs for older people. The sourcebook materials in these areas may provide you with a few new tips on program delivery possibilities, additional information resources to obtain, and some issues to consider when developing programs for special population groups, such as diabetics, or those who are non-ambulatory.

You can use the sourcebook as your *Joy of Cooking* for health promotion with older people; it will provide you with the information and resources you need to get started. You may want to use the manual as your own training course in health promotion with older adults or as training materials for the people with whom you will be developing and delivering the program. Finally, depending upon your participant population, you may want to use all or parts of the sourcebook as the basic text for the program that is actually delivered.

WHO WILL FIND THIS USEFUL?

Because older people live, work, and learn in a variety of settings, *A Healthy Old Age* is designed to be of use to a diverse group of trainers and providers from hospitals, nursing homes, senior centers, continuing education centers, and congregate living facilities.

Often, health services professionals lack sufficient information about the concerns and central issues of the aging network in relation to program design and development. The same information gap is true for many aging network personnel when it comes to understanding some of the basics of health promotion programs. The sourcebook addresses this deficit by providing a basic core of information which is relevant to both. It contains material for those who have experience in health promotion activities as well as those who are just beginning to gather resources and develop programs. Older and younger people alike who are interested in obtaining some basic information about health promotion and promoting their own health will also find the material relevant and useful.

Service providers and program planners, paid and volunteer, will find that the sourcebook is developed to address many of their questions and concerns about health promotion with the elderly:

1. What is health promotion?
2. What does it mean for older people?
3. What constitutes a health promotion program?
4. What are some examples of health promoting activities?
5. How do I structure such activities?
6. Where do I go for more information about resources?
7. How do I go about developing a health promotion program in my setting?

With tailoring and adaptation, both the content and the educational approach presented are appropriate to many populations: the well-elderly; the chronically impaired; the institutionalized; and intergenerational groups, too. The ways in which the sourcebook is meaningful and useful depends upon the population, the setting, and available resources, and most of all, the way in which you, the reader, integrate and interpret this information.

ORGANIZATION

A Healthy Old Age is divided into two parts—Background Information and Resources. The Background Information will be most useful in the initial stages of planning, marketing, and designing a health promotion program. This section provides some answers to the what, why, and how questions that emerge during the process of developing or expanding a program.

In "Wellness in Old Age" a rationale for health promotion, especially with older people in mind, is provided along with discussion of the range of possible health promoting activities. A description of one comprehensive health promotion program, the Wallingford Wellness Project, is presented to provide evidence that these kinds of programs are possible, appropriate and effective in improving the quality of life of participants. The Wellness Project experience provided the basis for the materials and educational approach taken in the sourcebook.

"Program Development Guidelines" addresses such issues as needs assessment, site selection, recruitment, and program evaluation . . . the organizational considerations that can make or break a program at the outset. In addition to discussing some tried and true principles of planning and program development, we provide in the "Dilemmas" section some of our experiences and knowledge about pitfalls and pratfalls that you may be able to avoid or overcome.

In "Changing Lifestyle: Learning Health Promotion Content" a participative learning model is presented as a good educational approach for health promotion programs. This model, which is compatible with the principles of health education, recognizes the critical value of participants' common sense expertise, and lends itself to building the interpersonal and program support that is needed for making and maintaining health promoting changes in behavior.

The importance of a comprehensive approach to promoting health is discussed in the "Pillars of Health Promotion." The core content areas of Nutrition, Physical Fitness, Stress Management, Personal and Community Self-Help, and Common Health Concerns are defined and examined in the context of a health promotion program.

The Resources section provides specific information on actual program delivery. Several different strategies for program format are presented in the "Variations on a Basic Theme" section. Each of the approaches discussed is compatible with the basic participatory model presented in Section One, but each involves the participant population in a different way.

"Tips for Facilitators" provides suggestions for implementing the educational approach chosen and the "Class Outlines" provides four curriculum modules for each of the core content areas. Depending upon your program needs and resources, these modules can be lifted "right out of the book" and implemented as your wellness program. They can also be used as a starting place for planning a series of classes or to supplement an existing program.

The "Common Health Concerns" section includes information and resources on health topics of particular concern to an elderly population. This section can provide the material for a lecture or discussion series which could be implemented with minimal staff and material resources. It also provides an "interest list" from which your program participants can choose topics most relevant to them.

The "References" section provides you with sources of additional discussion about each of the subject areas. For more definitive documentation of approach to health promotion you should consult these references. And for a detailed explanation of the Wallingford Wellness Project, its program and the associated research you should refer to the publications and reports of the project listed here.

We hope that *A Healthy Old Age* as a whole provides you with a foundation upon which to build, adapt, or expand your own program. The materials in it have been evaluated and re-evaluated by staff, by technical advisors, and most importantly by older persons who have participated in the design, development, and ongoing delivery of community-based wellness programs.

Some of the material, some of the techniques, and even the overall approach to comprehensive health promotion programming for older adults as presented in *A Healthy Old Age* have been questioned by some of those who have examined this sourcebook. While not all of the material will be effective all of the time with all older people . . . we have found that most of the information, most of the experiential techniques, and the overall participatory learning model is effective and enjoyable most of the time with many older people. There is no doubt that the basic approach is one which challenges the individual to develop, grow and change to achieve his or her optimal level of health and well-being.

For those with impaired physical or mental capacity, adaptations may be necessary and appropriate. The fact that the participatory learning model and the comprehensive four pillared approach to health promotion can be adapted readily to meet the needs and interests of different population groups suggests the power of the model. Already, it has been used with many different groups in many settings with results which please both participants and program developers. We recommend that if you are sceptical about the appropriateness of this approach to health promotion with older people, ask around. Find an individual or an agency that has tried using the materials and the participatory learning model of *A Healthy Old Age* and ask what worked and what didn't. Learn from their experience and then try it yourself in your own setting. In this way, you will benefit from the evaluation that only experience provides and you, too, can contribute to the refinement, adaptation and improvement of the materials presented in *A Healthy Old Age*.

A Healthy Old Age
A Sourcebook for Health Promotion with Older Adults
Revised Edition

BACKGROUND

Wellness in Old Age

RATIONALE FOR HEALTH PROMOTION

Since the turn of the century, the elderly population of the United States has multiplied approximately eight times. In 1900, people over 65 were about 4.1% of the total population; today every tenth American is age sixty-five or older. This trend is the graying of America, and it is expected to continue for the foreseeable future (U.S. Senate, *Developments in Aging: 1978*, A Report to the Special Committee on Aging, p. xxiv).

The 24 million elders now living are not necessarily testimony to advances made in increasing the lifespan. Life expectancy at birth has increased from 47 years to approximately 72 years, primarily as a result of the virtual eradication of many infectious diseases such as tuberculosis, diphtheria, and smallpox. Less progress has been made in preventing or minimizing the consequences of chronic conditions and diseases such as high blood pressure or heart disease. Although many more people live to be 65, they only live an average of 3.7 years longer than their counterparts who reached that age in 1900.

Most of the diseases that afflict older persons are the so-called lifestyle diseases. . . conditions like lung cancer, diabetes, and heart disease, that are strongly linked to risk factors such as cigarette smoking, poor nutrition, lack of regular exercise, and chronic stress. Eighty percent of older people suffer from one or more of these lifestyle diseases and chronic conditions and about half of all older persons are somewhat limited in their daily activities.

Simple preventive measures can significantly reduce the negative impact that lifestyle diseases have on older adults. For example, many of the limitations on activity experienced by older persons stem from respiratory conditions such as bronchitis and emphysema. Reducing or eliminating cigarette smoking could provide direct and immediate benefits to these persons. And, in 1977, influenza and pneumonia was the fourth leading cause of death for older people. Many of these deaths could have been prevented by vaccinations and proper personal and medical care.

The Surgeon General in *Healthy People* (1979) suggested three powerful reasons for increasing disease prevention and health promotion activities for people of all ages:

> Prevention saves lives.
> Prevention improves the quality of life.
> Prevention can save dollars in the long run.

Examining these in relation to the older population suggests that health promoting activities with this group are critical. First, prevention saves lives; the highest rate of accidental death and injury occurs among the elderly. Their risk of fatal injury is almost twice that of adolescents and young adults. Relatively simple and inexpensive accident prevention measures such as illumination of walking surfaces, handrails, smoke detection systems, elimination of smoking in bed, and optimum vision and hearing care, can significantly reduce the risk of accidental injury and death for older persons (*Healthy People*).

Second, as the Surgeon General suggests, we have succeeded as a society in helping more people to live longer. We should now look at another goal: "a better, healthier life for older people." Often, the limitations placed upon the individual by chronic and acute conditions inhibits the older person's ability to create an old age lifestyle of their choosing. The long-term goal of health promotion and disease prevention programs for the over 65 population must not only be to achieve further increases in longevity, but also to allow each individual to seek an independent and rewarding life in old age, unlimited by many of the health problems that are within his or her capacity to control.

The third reason is based on economics. Many older persons live in poverty, and many depend upon a fixed income that is minimally responsive to changing economic conditions and the changing personal economic needs of later life. Using these limited funds for any health-related expenses that could be prevented or reduced can unnecessarily inhibit the individual's ability to do things that would enhance the quality of daily life. As a nation, we can ill afford to continue subsidizing sky-rocketing health care costs that could be reduced through individual and community efforts in prevention. The elderly constitute 10% of the population and account for 30% of health-related expenditures; they have health-related costs *over three times* that of the general public. Anything that can be done to reduce the need for these expenditures can benefit not only the individual, but society as a whole.

WHAT IS A HEALTH PROMOTION PROGRAM?

Health promotion has been described by Green as ". . . any combination of health education and related organizational, political, and economic changes conducive to health" (Lowe, 1980, p. 9). A health promotion program, then, is designed to improve the health and wellbeing of individuals and communities by providing people with the information, skill, services, and support needed to undertake and maintain positive lifestyle changes.

Health promotion programs recognize that individual health is influenced not only by voluntary personal behavior but also heredity and economics, and the social world in which we live.

At the community level, health promoting activities can be diverse, ranging from water fluoridation, blood pressure screening clinics, immunization programs, and comprehensive lifestyle change programs for individuals and groups. All of these programs recognize that education and environmental changes that promote health occur as part of a team effort between all members of the community. For example, disease prevention and health promotion efforts occur in partnership with health care providers through medical treatment and cure of disease, and individual and community health enhancement programs. These activities are designed to prevent or manage disease and disability, with the common goal of securing optimum health for the individual and the community.

Health promotion programs also recognize that individual self-responsibility plays a key role in promoting overall wellbeing. Personal health can be significantly improved through individual actions such as appropriate exercise, proper diet and stress management. Personal health is also enhanced by individual measures such as accident prevention in the home and in the community.

For individuals, health promoting activities are most successful when each person has access to relevant information and services, motivation to secure needed information and services, the skills and resources necessary to act on the information he or she has acquired, and the personal and community support for making and maintaining positive health behaviors.

Various aspects of community support are important in pursuing a healthy lifestyle. Family and friends who encourage and support positive behavior changes are one form of support. Friends and fellow citizens who work to reduce local air pollution, secure safe places to exercise, encourage local markets to provide healthy foods such as low-fat cheeses or fresher fruits and vegetables are another form of support for promoting healthier lifestyles. Working to increase the wellbeing of the environment also adds to individual wellness indirectly, in the long-run, by creating a safer and healthier living situation, and directly by providing the individual with an enhanced sense of contribution and self-esteem.

HEALTH PROMOTION WITH THE ELDERLY

Health promoting activities can be designed to address specific diseases or problems that are common to a particular population group. Most of the health-related activities currently taking place with older adults belong to this group. These activities address one particular problem, such as hypertension or cancer screening, or focus upon one aspect of behavior such as smoking cessation, or weight control. Programs such as these are essential to disease and health problem management. They reduce the individual's risk of poor health or unnecessarily rapid disease progression. They also provide a starting point for developing more encompassing health promotion programs.

For older adults, the Surgeon General's Report, *Healthy People*, identifies an overall goal for health promotion programs:

> To improve the health and quality of life for older adults, and by 1990, to reduce the average annual number of days of restricted activity due to acute and chronic conditions by 20% to fewer than 30 days per year for people aged 65 and older. (p. 71)

As the report suggests, ". . . staying healthy is an ever changing task . . . at each stage of life different steps can be taken to maximize wellbeing" (p. 16).

A health promotion program which incorporates the basic principles of self-responsibility for health behaviors and support for making and maintaining health promoting behavior changes will be most beneficial if it addresses a range of everyday health habits. For example, reducing the consequences or the risk of many chronic diseases is best accomplished by a lifestyle which includes proper nutrition, sufficient exercise, well-managed or minimized stress, and a personal and community environment which is conducive to supporting activity in these areas.

Developing or expanding a program to incorporate just these four areas, nutrition, exercise, stress management, and personal and community self-help, is a big job. Securing the needed resources, material and personnel, enlisting the support of the community, and recruiting the participants for the program is a challenge. However, the potential payoff is comparably great. The cumulative benefits for participants, including staff, of making behavior changes in these basic areas are well worth the efforts. Some of the benefits are fairly obvious;

—reduced risk of illness and discomfort
—more effective management of existing chronic conditions
—more appropriate use of the medical care system and reduced medical expenditures.

Other benefits are less obvious, but perhaps even more important:

—an increased sense of control and independence
—an increased sense of strength and wellbeing
—an increased sense of being a valuable participant in community life.

THE WALLINGFORD WELLNESS PROJECT

How do the Surgeon General's goal and the associated benefits translate into programming reality? Available research shows the importance and benefits of health promotion programs which have focused on younger persons. However, comprehensive lifestyle change programs with older people have been few; documented and evaluated health promotion programs with older people are even scarcer. The Wallingford Wellness Project (WWP) provides one example of a community-based comprehensive health promotion program which focuses on lifestyle changes with older people. The WWP provides the basis for the approach to health promotion with older people which is proposed in *A Healthy Old Age*. The WWP has demonstrated that a comprehensive health promotion program with older people as participants and leaders is effective, appropriate and exciting.

The Wallingford Wellness Project began in 1979 as a model project of the University of Washington School of Social Work and Senior Services and Centers of King County, Inc., funded by the U.S. Administration on Aging. The purpose of the project is to develop, evaluate and disseminate a health promotion program with older persons which focuses on nutrition, exercise, stress management, and environmental awareness—everyday life concerns that directly affect the health and wellbeing of people of all ages. Several features of the WWP are distinctive:

1. It is participant-oriented. Participants and potential participants, as well as staff and service providers, have been involved in all phases of the project, from developing the original recruitment plan, to taking part in the delivery and evaluation of the training program.
2. The long-range goal of the training program is individual and community empowerment. The focus is on helping participants to assume maximum self-responsibility for overall wellbeing. By emphasizing the acquisition and improvement of the knowledge and skills needed for healthy everyday living, participants are able to use these tools to all aspects of their lives, not just those immediately related to health.
3. It is intergenerational. About one-quarter of the participants in the WWP have been under 60. The ages of participants to date have ranged from 13 to 84.
4. The educational component emphasizes common sense: health-related information and skills that are well within the grasp of most people. The need for extensive involvement of "experts" in the training program is minimized. Focusing upon basic health practices and relying on the reservoir of information and expertise of the participants allows the costly time of specialists to be used sparingly, as a supplement and in response to participant's interest in pursuing specific topics in greater detail.
5. Training in communications skills and assertiveness is a key program component. Originally conceived as an "Environmental Awareness/Assertiveness" course, this part of the program evolved to incorporate and emphasize a variety of

communication skills. The application of these skills is diverse, including, for example, improved relationships and information exchanges with health care providers, friends and family members, working with others to secure desired products at local shopping places, and organizing to influence local legislation to enhance traffic safety provisions for older people.

Approximately three hundred older persons have completed the training program so far. The first two groups of participants were led and staffed primarily by professional social workers and social work interns. The third and subsequent groups of participants has been staffed primarily by participant "graduates" and coordinated by social work students. Support services, clerical work, and the day-to-day "nitty gritty" of on-site administration have been provided primarily by one professional during the first two training groups. During the third series of classes, volunteers (also program graduates) have assumed these responsibilities.

The Wallingford Senior Center, where the project is located, has gradually incorporated the Wellness Project into its own program and has allocated staff time for coordination of future efforts. Program graduates are staffing the fourth series of wellness classes, a Speaker's Bureau, an Advisory Committee, and ongoing support groups which provide graduates with a forum for working to maintain and extend the healthful changes which they began in the core training program.

The core training program has changed considerably over the past three years. Originally, the group met for twenty-four weeks of class, three hours per week. During the three hour session, participants attended two classes. One class was always Environmental Awareness (Personal and Community Self-Help), and the other class was an eight week session of Nutrition, Stress Management, and Physical Fitness. The second group met for twenty-one weeks, following the same pattern, and the third group met for fifteen weeks. The program no doubt will continue to change in the future to meet participant and staff needs. The abbreviation of the program length was in response to available staff resources (paid and volunteer) and participant feedback. Developing a program which is reasonable and manageable in terms of the needs of ongoing volunteers has been a major concern of the project.

The materials presented in this sourcebook represent the best and most current experience of staff and participants in the Wallingford Wellness Project. Some of the materials, the Common Health Concerns fact sheets for example, were developed in response to interests expressed by participants. During the core training program, these topics were often presented by outside resource people, an example of the focused use of "expert" time.

Research on this project, its efficiency, effectiveness of promoting healthful behavior changes, and the appropriateness and responsiveness of the model to an older population has been extensive. The quantitative research used a quasi-experimental design and examined measures of social and mental health, health knowledge and health behaviors in the four program areas, and morbidity and health services utilization rates.

The program evaluation focused upon the actual program delivery model and the perceptions of participants about program content, presentations, and staff effectiveness. Results of these evaluations are available, as well as detailed descriptions of program components and the educational materials (FallCreek, 1982, The WWP Staff Team, 1982).

The best evidence of success is available from the actual program participants:

Ben Tappe, age 74: The proof of the pudding is in the eating. The Bible says when you live three score and ten you are then on borrowed time. I attended twenty-four weeks of lessons and talks about health and nutrition and I have set *my* goal to reach ninety.

Bonnie Dewey, age 57: Participating in the Wellness Project has been a very enriching and fulfilling experience for me. It has helped me to learn more self-confidence and to accept changes in my life much easier. I am now able to do things with ease that I would have found very difficult before. Each class contributed to this, and added its needed part to the whole well-rounded program. Making new friends who were also interested in accomplishing helpful changes in their lives was another plus. I also cannot praise the staff enough for their very loving and inspiring help.

Erica Duringer, age 73: I have been able to make changes in my lifestyle which didn't happen by reading books or watching TV. And why did it work? Learning in a group helps, devoted and capable teachers, giving feedback, and receiving feedback. A valuable experience and I became a healthier and happier person being more aware about the needs of others.

Virginia Smith, age 68: I knew about these things before, but this gave me a boost. It was good to know other people are involved and it also helped me with keeping my husband's blood pressure down. I'm also glad the government is taking a stand on this sort of thing. I'm glad the word is getting around.

Elva Holland, age 63: The Wallingford Wellness Project has brought me to a significant awakening. Something *can* be done about one's state of health and it is *not* too late!

I fought the idea of four subjects and several months of classes with my already busy schedule and therefore asked if I could please audit the nutrition segment. I was told they considered it preferable to attend all four segments since they were closely correlated. Very soon I realized the wisdom of this and as the classes progressed I found it easy to eliminate schedule conflicts in favor of the friendly, caring, and sharing atmosphere of learning with the WWP. I didn't want to miss a single session!

In the Nutrition class I learned to reduce salt, fats and sugars and this, plus more regular exercise, has helped me lower my blood pressure and lose a little weight. In the Stress class I learned techniques to relax myself to sleep after a stressful day.

In Environmental Awareness, the hours of the Puget Consumer's Food Co-op, the Urban Homes League, and the Arboretum Tour (led by another participant, Della Patch) all made me aware of the need for a healthy environment.

It has been one of the more enriching experiences of my life to be among the first "grads" of the Wallingford Wellness Project and to be able to encourage others to seek the many benefits of this outstandingly valuable program. The friendships that developed from our class appear to be lasting ones. I am indeed grateful!

Jean Newman, age 72: These are some of the results of my participation in the project: lost several pounds, reduced blood pressure, reduced food budget, have more energy, made excellent recovery from hip surgery, learned it was okay to say "no" and not feel guilty, I now recycle cans and bottles, reuse items like paper and plastic bags, and read the labels on food products more carefully. I am proud to be a pioneer in this project!

These participant testimonials are supported in several ways by the research and program evaluation data that were collected during the first two years of the project (Lalonde and FallCreek, 1981). Interest and excitement about the project and its educational model extends beyond the boundaries of the Wallingford neighborhood in Seattle where the project is located. Hundreds of requests for information, research findings, and educational materials have been received from across the nation. Perhaps the best evidence, though, of the value of this model for health promotion with older people is found in the commitment of the project's participants to keep the program alive and well without the federal funding which initiated and supported the original effort. Graduates of the training program are now leading classes themselves, not only in the Wallingford Senior Center, but in other places where older people congregate—housing projects, highrise buildings, and neighborhood centers.

Health promotion programs with the elderly are an important addition to the existing health care system. Health promotion assumes and reinforces the premise that older individuals are capable of learning about and promoting their own health given access to the skills and information needed. It enhances the capability of older adults to maximize their independence and improve the quality of their lives. This makes health promotion a powerful concept well worth the time, effort, and commitment needed to work toward a healthy old age.

Program Development Guidelines

In suggesting guidelines for developing a health promotion program, we recognize that each setting has unique people, conditions, and resources. The service goals which influence the decision to try a health promotion approach are also unique. Therefore, our intent is to provide a general orientation toward delivering health promotion programs and guidelines for adapting the information presented in this manual to your setting.

To ensure a common frame of reference, "community" is defined as any group of people living together or having close interaction. "Community" includes all the people—older adults, service providers, support people, administrators, family, and friends—who interact in any given setting. For example, the community might be composed of people maintaining their own residences in a larger community, people who live in a nursing facility, people who share activities at a senior center, or people who come to a hospital or agency for health care. "Service provider" is any person who plays a role in social or health care services with older adults. Health promotion programs can be started by activity directors; nurses; social workers; occupational therapists; aides, doctors; nutritionists; health educators; paraprofessionals; volunteers; that is, by any member of the community.

Adopting health promotion strategies—like health itself—is a continuum. Initial acceptance may be slow. The same techniques which help people take control of their health can maintain your enthusiasm as you start a new program: start where you are, make the easiest changes first, plan ways to check your progress and make revisions, and allow for setbacks. Most of you will be pioneers of this approach in your setting. We encourage you to take a risk and try a participation-oriented model of service delivery. Based on the experience of existing health promotion programs with older adults, the guidelines and tasks that follow will help you get started in your community.

GUIDELINES FOR ADOPTING HEALTH PROMOTION SERVICES

This chapter addresses the practical question, "How do I begin to apply this information in my setting?" Our goal is to review components of program development and service delivery that may deserve consideration in your setting. Given the diversity of possible settings, you will have considerations which are not covered here. Similarly, in an attempt to be comprehensive, we will cover topics that will not concern every reader.

Each guideline listed may play a role in establishing or expanding a successful program. Some guidelines overlap; all are interrelated. For each guideline, tasks are suggested to provide direction for developing a program within a given community framework. This format is not necessarily chronological. Priorities and scheduling of tasks will be determined by interests and resources unique to your community.

GUIDELINES

Program Rationale

—Identify older adults' health concerns and motivation to improve health.
—Identify potential resources in the community.
—Prepare recommendations for a health promotion program in your setting.

Community Consensus

—Identify potential participants and advocates in the community.
—Attain endorsement from the board members or agency administrators in your community.
—Build consensus regarding the health promotion model among program staff.

Organizational Structure

—Develop avenues for collaboration and communication.
—Specify involvement of staff members in the health promotion program.

Site Selection

—Identify a location for conducting the program.

Staff Skill Development

—Clarify health values held by staff members.
—Train staff to lead classes.
—Build into the program some means of systematically rewarding staff members for leading healthy lives.

Participant Recruitment

—Develop a recruitment plan.
—Recruit participants.
—Recruit with the future in mind.

Program Evaluation

—Determine the information needs in your community.
—Develop plans for securing needed information over the course of the program.

Fitting Your Population

—Common Concerns
—Sample planning considerations

Dilemmas

—Not enough planning time
—Delivering a prescription
—Turf, investment, and cultism

PROGRAM RATIONALE

An assessment of older adults' health-related interests, attitudes, skills, and behaviors can be the basis for your health promotion efforts. While written documents may reveal health problems of older adults, less formal procedures, such as conversations or interviews, are more likely to uncover how older adults maintain good health and what attitudes they hold regarding aging and health. Discussions with groups of older persons may help you identify the special interests and health beliefs common to your target population. For example, if a large number of older adults in your setting rely on traditional, culturally specific health care providers (e.g., a curandero, or an herbalist), you should plan to include those providers in your planning process, as well as recognize in your program design the nature and significance of the health practices they advocate.

Community health care providers can contribute valuable information on the type of health care services currently sought by older adults. They can also indicate gaps in service that might be filled by a health promotion program and help tie the program into the overall service network available for older adults.

A thorough analysis and understanding of the health concerns of your community and available resources will provide a solid rationale for a new program. Match your immediate goals with resources readily available. Keeping the resources of the community in sight while making recommendations will reassure members that the program will succeed.

Identify Older Adults' Health Concerns and Motivation to Improve Health

Assess the attitudes and concerns of older adults and their service providers regarding health and aging.

—surveys
—interviews
—questionnaires
—other

Review community reports on previous health education or promotion efforts.

—agency records
—community newsletters
—information from participants
—program evaluations
—other

Learn how community members—including administrators, service providers, and older adults—communicate, make decisions, and respond to health concerns and available services.

—interviews
—informal conversations
—other
—staff meetings
—agency policies

Summarize the health concerns of older adults in the community and the past and current approaches to providing health care services. As the need for a health promotion program is based in the older population, their interests and preferences should guide the planning process.

Identify Potential Resources in the Community

Summarize the available resources.

—equipment
—human resources
—personnel
—space
—materials (pamphlets, handouts, etc.)

Contact community members that might support or contribute to the program.

—members of the clergy
—editors of newspapers
—hospital administrators
—service staff (cooks, nurses, janitors)
—social workers
—members of the board of directors
—hospital dietitians
—nursing home administrators
—community council members
—volunteer organizations
—alternative health care providers
—educators
—lobbyists or special interest groups

Investigate available funding within and outside your community.

—fee for service
—local businesses
—large corporations
—volunteers
—government resources
—local foundations
—donations
—local government agencies
—fundraising events
—other

Summarize available human resources in the community and indicate what they can contribute to a health promotion program.

—professional skills (paid or volunteer)
—space for the program
—materials
—equipment
—media exposure

—support services
—other
—participants

Develop a budget for your proposed program that covers essential costs.

—personnel (including expense for volunteers)
—consultation fees
—educational materials
—space
—supplies
—insurance
—other

Prepare Recommendations for a Health Promotion Program in Your Setting

Describe the target population for the program.

—age
—ethnicity
—economic access to services and programs
—health status
—mobility

List the specific health care needs of your target population.

—exercise programs
—stress management
—high blood pressure management
—other
—nutrition information
—alcohol treatment
—diabetes control
—home accident prevention

Outline program goals and objectives for your health promotion program. Recommend the learning objectives and teaching methods to be used. Base these goals, objectives, and methods on the needs, skills, and interests of both the potential participants and the program leaders and service providers.

COMMUNITY CONSENSUS

Involve community members in program development—their participation and enthusiasm will play a vital role in community acceptance. The first step is to introduce the principles of health promotion. Your role is to facilitate change in the service orientations held by community members and in the service itself and to recognize and respond to the particular health beliefs and practices of the target population and those who provide their health care services.

Your role is also to incorporate appropriate changes into your vision for health promotion. Potential participants may prefer an emphasis on oral rather than written communication. If many participants are also Spanish-speaking, for example, you will need to incorporate the services of Spanish-speaking staff, paid or volunteer, into the program delivery plan.

Draw on the resources in this manual in your campaign to create a shared vision. Once people are interested and asking for more information, you are ready to begin building consensus. Share health promotion information that is relevant to individual needs so people can begin to experience the personal benefits of health promotion.

Identify Potential Participants and Advocates in the Community

Contact respected people who work with elderly people and might support or participate in the program.

Contact leaders in the health service community who might support or contribute to the program.

Contact leaders among the community's social service providers who might refer or recruit participants.

Contact leaders in the spiritual, political or religious community who might give approval or support for participation to their members.

Obtain Endorsement from Board Members or Agency Administrators in Your Community

After making individual contacts, call a small group meeting to review the health promotion program recommendations with members of the community who influence service policy.

—executive director of an agency
—activities coordinator
—director of social services
—hospital administrator
—head of the nursing department
—other

Solicit comments on the specific recommendations for the health promotion program.

Request endorsement of the proposed program.

Build Consensus Regarding the Health Promotion Model Among Program Staff

Familiarize staff members with the health concerns of the target population.

Explain the health promotion model. Solicit their feedback.

—handouts
—in-service training
—demonstrations
—workshops
—guest speakers or consultants
—other

ORGANIZATIONAL STRUCTURE

Health promotion programs can develop in a variety of organizational structures. Consequently, we highlight two of the elements that may facilitate implementation of a program in any setting—effective communication and staff job descriptions. Communication will help maintain the collaborative approach to service delivery and specific job descriptions will help clarify any change in service orientation required by the new program.

We recognize that a sole service provider may want to add health promotion components to ongoing services. For the person working alone, reading materials suggested in this manual and seeking people in the larger community who value the health promotion model can be a source of inspiration, purpose, and skill development. Program participants can also provide a wealth of ideas and experience.

Develop Avenues for Collaboration and Communication

Specify the key people within your setting responsible for coordinating service delivery.

—program coordinator
—social services department
—other
—activities department
—nursing staff

Organize a steering committee composed of members of the administrative staff and of the health promotion program staff. Responsibilities of the committee might include:

—publicity releases
—media coverage
—program evaluation
—recruitment procedures
—funding actions
—future directions

Organize a program advisory committee comprised of program participants and staff members to work cooperatively on developing the program format and to plan additional activities such as the following:

—a social gathering for staff and participants
—an advocacy program for older people
—recruitment of new participants
—shared leadership of program classes among staff and elderly
—revision of program activities

Specify Involvement of Staff Members in the Health Promotion Program

Define staff member roles as participant-leaders in the program—emphasize that they will facilitate the transfer of information and self-help skills.

Share responsibility for developing program content with staff members.

Collaborate with staff members in developing job descriptions that reflect participation in the program.

—social worker
—health educator
—program secretary
—occupational therapist
—activities leader
—other

Specify criteria for personnel evaluation.

—quantity of work performed
—management of program resources
—performance as a program member, communication skills
—ability to be supportive
—ability to counter ageism, racism, and sexism

SITE SELECTION

The site chosen for the health promotion program should be one that already attracts or will attract the target population. The location should be one where participants can practice health-related changes. If your program includes exercise, then you will need a room large enough for an exercise class. If you want to include a nutrition component, then it would be helpful to have a kitchen. The scope of your program will determine the site requirements. Many sites may already offer health-related activities and can more readily adopt a health promotion approach.

Identify a Location for Conducting the Program

Select a location that attracts or will attract your target population.

Select a location where people can practice health-related activities.

Anchor the location of the program within the community by planning or developing publicity strategies.

—public meetings
—telephone contacts
—radio or television announcements
—newspaper articles
—other
—press releases
—senior centers
—bulletin board posters
—brochures
—churches
—nutrition sites

STAFF SKILL DEVELOPMENT

Our basic premise is that service providers can best facilitate change for the elderly when providers themselves are actively engaged in improving their own health and well-being. By example, staff encourage participants to take greater responsibility for their individual health, and demonstrate the benefits of health enhancing changes.

People new to health promotion may need training and practice in specific leadership skills that facilitate self-selected change by older adults. Learning to apply health promotion skills in our own lives is the first step towards effective leadership. The staff does not have to be a model of health, but rather a model of the step-by-step improvement that is possible. In fact, staff sharing personal challenges and frustrations in health promotion helps to make the program believable and possible for participants and other staff members. Staff experience the learning process with participants.

Clarify Health Values Held by Staff Members

Guide staff in assessing their own health values and the health expectations they hold for older people.

—values clarification exercise
—questionnaire
—discussion groups
—other

Train Staff to Lead Core Classes

Trainers model for the staff members the specific skills required by the program content—develop a series of mini-presentations for the staff. Have staff members practice these skills as a group and exchange comments with each other.

—small group facilitation skills
—generating participation of the elderly
—other
—communication skills
—organizational skills

Guide staff in developing class agendas.

—refer them to the curriculum development sections in this manual
—provide examples of appropriate agendas

Build into the Program Some Means of Systematically Rewarding Staff for Healthy Lifestyle Efforts

Develop a method of staff collaboration.

—regular meetings
—routing slips
—bulletins
—informational memos
—posted minutes of meetings
—wellness days

Develop a system of regular evaluation with staff members of self-selected leadership goals.

Make "health days" an alternative to "sick days" for staff members. For example, a staff member who is not sick over a two month period would accrue a "health day" of vacation.

Encourage staff members to support each other in developing more health-oriented habits and lifestyles.

PARTICIPANT RECRUITMENT

Collaborating with community members is a key factor in initially recruiting program participants. The following steps help build community involvement. Once the program has been tested in the community, former participants can recruit new people to the program.

Develop a Recruitment Plan

Develop a list of potential screening criteria for program participants.

—health status
—availability
—commitment
—age
—income
—other

Determine how many people the program should aim to serve.

Develop a one-page handout that outlines recruitment goals and specifies questions you have about how to reach these goals. (See Figure 1.)

—program sponsorship
—target population
—attributes of desired population
—philosophy of program
—goals of program
—recruitment questions

Schedule a meeting to solicit input and advice on recruiting procedures from service providers and active, interested members of the older adult community. Use the recruitment handout to focus the discussion.

Plan recruitment strategies and materials based on these recommendations generated at the meeting.

Notify the people who attended the recruitment meeting of your plans and thank them for their assistance, and offer them opportunities for continuing participation in recruitment and other aspects of program development.

Recruit Participants

Publicize the program and request the community's help with recruitment.

—program brochure
—media announcements
—press releases
—other
—contact service providers
—contact community leaders

Develop a procedure for responding to applicants for the program.

—information to be given to applicant
—information needed from applicant
—how to answer "tough" questions
—how to record contact
—other

Prepare a screening questionnaire to determine, if necessary, eligibility of potential participants, and to identify conditions which you will need to take into account in designing the program. (See Figure 2.)

—name, address, telephone number
—prescribed medications, over-the-counter medications
—relevant medical history
—mobility, physical ability
—other

Schedule open public meetings to recruit participants at times and locations that will attract older people in the community.

—activities room
—senior center
—recreation center
—lunch site
—church meeting room
—other

Prepare a one-page handout to be distributed during meetings with potential participants.

—program structure
—program administration
—fee
—other
—program sponsorship
—summary of program philosophy
—brief description of classes

Present a mini-version of your program at the public meeting to build interest and trust.

—group exercise demonstration
—nutritious refreshments
—relaxation technique
—risk factor analysis
—other

Distribute one-page handout and screening questionnaire during the public meeting and help people complete these forms before they leave. If necessary, follow screening meeting with a phone call to arrange time for a more comprehensive assessment of health history, attitudes, and behavior.

Schedule attendance at an introductory workshop.

Schedule an introductory workshop to present a preview of the classes that will be included in the program, establish commitment to the program, and have participants sign up for their first class.

—demonstrate program activities
—introduce staff and participants

—establish location, meeting times, important telephone numbers
—review program content
—schedule participation
—other

Recruit with the Future in Mind

Keep good records. Potential participants may have a scheduling conflict that prevents enrollment in an initial program. They can be contacted again later. Many people may adopt a wait and see attitude. They are good candidates for future programs.

Evaluate what recruitment strategies were most effective. Ask new program enrollees how they heard about the program or what initially attracted their interest.

FIGURE 1: First Recruitment Meeting Handout (Sample)

The __________ (name of department, service, agency, or setting sponsoring the program) will be developing a number of classes dealing with health promotion. These classes will involve __________ (specify the population that will be recruited) in learning how exercising, controlling stress, and eating healthful foods can improve one's health and prevent illness and dependency. These classes will be designed to build on the positive health and living habits of participants, as well as to generate motivation, support, and tools for choosing and maintaining individually healthier lifestyles. The first classes on physical fitness, stress management, nutrition, and personal and community self-help will engage participants and staff in cooperatively working out a program that best suits our setting and people.

We are looking for people to participate in this project. We would like to find __________ (here specify the number, age, and other important attributes of desired participants). To find these people we need your help in answering the following questions:

1. How do we locate people who would be interested in this program? Whom do we talk to? What information do we look for?
2. Who are the service providers to potential participants? Of these service providers, who might support this project? What information can we supply to increase support for the project?

3. What will participants want to know about the project before they will get involved? What can we tell them that will lessen any fears and generate enthusiasm?
4. How can we best communicate the purpose of this new program? What communication tools should we develop, i.e., brochures, workshops, in-service training, etc.?
5. How should we approach people, i.e., go door to door, telephone, existing service providers, etc.?

FIGURE 2

Sample

Screening Questionnaire

Please indicate, with a check mark, which, if any, of the following physical activities are difficult for you:

	CAN DO WITHOUT DIFFICULTY	SOMEWHAT DIFFICULT BUT CAN DO WITHOUT ASSISTANCE	DIFFICULT AND NEED ASSISTANCE FROM ANOTHER PERSON	For Proj. Use Only
1. Walking up & down stairs				___ ___
2. Picking up objects from the floor				___ ___
3. Getting in & out of a chair, bath-tub or car				___ ___
4. Reaching for ob-jects above shoulder level				___ ___
5. Cutting Your Toenails				___ ___

Do you have, or do you suspect you may have, any of the following conditions? ___ ___

___HEART PROBLEMS ___LUNG DISEASE ___RHEUMATOID ARTHRITIS

___THYROID DISEASE ___SEVERE DEPRESSION ___CHRONIC INFECTION

___DIABETES ___OTHER MEDICAL PROBLEMS (LIST): ____________

____ ___ ___

Are you taking any prescription medications? (please list): ____________

____ ___ ___

Please indicate any other medical information you think is important for us to know: ______________________________ ___ ___

NAME: ____________________ PHONE NUMBER: ____________

ADDRESS: ________________ BIRTHDATE: ______________ ___ ___

Day Month Year

____

PROGRAM EVALUATION

For many service providers, starting a health promotion program with older people will be exploratory. "Will this type of program 'work' with my population?" "Will people participate?" "Is it relevant?" Questions such as these indicate that service providers want a program that responds to the needs in their community; they want a program that can be sensitive and appropriate for a specific population. Guided by these concerns, the health promotion model described here asks service providers to consider program development and evaluation.

By necessity, a new program is experimental. Even the best curriculum is likely to change after its first test in application. This health promotion educational model encourages adaptability by integrating evaluation into program delivery. As people take part in the program, they provide information on what works for them and what doesn't. Service providers can then adjust the program based on the comments of past participants. For example, a simple evaluation form administered at the end of each class provides information on ongoing effectiveness.

Over the course of the program, participants will change, as will the needs and interests of the group. In the Wallingford Wellness Project, participants in the second training group had significantly more chronic conditions and disabilities than the initial participant pioneers. More members of low-income and/or ethnic groups may become participants, requiring a shift in content or a difference in teaching method. The evaluation process helps to keep group leaders and program planners on their toes and sensitive to the changing needs of the participants.

In addition to making the program more relevant to the community through participant evaluations, service providers will need information to reasonably evaluate the total service. "Have people increased health behaviors and attitudes?" "What kind of people were attracted to the program?" "Why did people drop out?" "Were older people able to use health care services in the community more effectively?" "How many staff hours were required to deliver the program?" Answers to these questions can help the service provider assess the program's objectives, methods, and services as well as provide a basis for future funding requests.

Program evaluation can include both subjective (e.g., self-report measures, interviews) and objective (e.g., blood pressure test, diet survey, standardized instruments) sources of information. This information will indicate the outcome effect of the program and provide information for revision and development. By generating information for improving the program within the community, service providers develop sources of support for health promotion. This can have an important long-term benefit since it is the support of the programs and people of the community at large that will help the program participants maintain healthy behaviors.

Determine the Information Needs in Your Community

List your commitments regarding program operation.

—grant stipulations
—budget constraints
—other
—audit procedures
—federal or state reimbursement requirements

List the program resources that require monitoring.

—participant fees
—fiscal resources
—human resources (e.g., staff, volunteers)
—physical resources
—(e.g., equipment, space)
—other

List the information about participants that requires monitoring.

—attendance
—demographic (e.g., age, sex, race, education, income, employment)
—attrition
—use of other health care services
—other

List the outcome variables that require monitoring.

—overall level of health
—mental health status
—health problems
—days of limited activity
—heart attack/stroke risk
—health care utilization
—other
—change in health behaviors
—social health status
—self-rated health
—morbidity rates
—health knowledge
—maintenance of change
—use of medications

Develop Plans for Monitoring the Selected Variables

Select methods of monitoring the program operations.

—program advisory committee
—regular consultation and supervision
—regular review
—account procedures
—reporting mechanisms
—other

Select methods of monitoring program resources.

—staff time sheets
—daily logs
—accounting procedures
—other

Select methods of monitoring participant use of the health promotion program.

—attendance sheets
—demographic questionnaire
—dropout questionnaire
—other

Select methods of monitoring participant satisfaction with the curriculum and program delivery.

—logs
—journals
—other
—consumer evaluation forms
—followup questionnaire

Select methods of monitoring selected outcome factors.

—diet survey
—health knowledge questionnaire
—health status questionnaire
—other
—physiological measures (e.g., blood pressure rating, weight loss)
—social health questionnaire
—utilization of health-care services questionnaire

FITTING YOUR POPULATION

Trainers and program developers need to be sensitized to some of the issues in planning and implementing health promotion programs with any specific target population, such as ethnic minority, or institutionalized elders. First, common concerns and barriers to making health promotion programs accessible are identified and briefly discussed. Second, samples are provided to guide the trainer or service provider in considering the kinds of information which may be pertinent to designing health promotion programs for several of the populations currently underserved.

While general trends, population characteristics, or concerns can be identified, the diversity of the individuals in the target population should be the starting point for program planning efforts. The differences between individuals within any specific population group, let alone among several groups, demand that the program planner and the service provider rely primarily upon information obtained from prospective program participants or indigenous community leaders.

It is very important in designing, developing and implementing programs to begin where the participants currently are, in terms of knowledge and skills. This also applies to motivating participants to begin and to continue making health promoting changes in daily habits. For example, strengthening exercises may enable a participant to walk to the grocery store and carry home a bag of groceries instead of hiring a taxi or depending upon a ride, a stress management technique may enable a participant to go to sleep more easily and sleep more soundly through the night. Range of motion and stretching exercises for the arms and hands may make it possible for the participant to button an heretofore inaccessible button or reach a previously inaccessible shelf. It is the responsibility of the program planner to assess the needs and interests of participants and design the program accordingly.

The importance of participant input and cultural sensitivity and awareness extends across ethnic, class income, geographical, and disability boundaries. Working with persons and groups who have experienced a distinctive history, such as racial discrimination, who have English as a second language, or who are isolated from mainstream human services requires an extra effort from program developers in reducing barriers to access and increasing use of health promotion programs.

The populations considered here include ethnic minority, chronically disabled, institutionalized, and low-income older adults. These populations overlap. The population of low-income elderly is clearly the largest group and, conceptually, it incorporates most other underserved groups. Another similar grouping can be identified: most underserved groups tend to be in poorer health than the rest of the older population. Thus, program design and development particularly with these populations should recognize and respond to inadequate financial resources and reduced overall level of health. For example, a physical fitness component which involves a jogging program and expensive jogging shoes is inappropriate. In fact, a program design which fails to take into account the health and economic limits of the target population actually may be counterproductive, resulting in physiological or psychological injury and/or inappropriate use of available resources. Programs must be designed on the basis of the resources available to the target population and specific health needs and interests.

Another set of present potential barriers to health promotion are cultural and demographic characteristics of the target population. The nature of the characteristics or conditions vary from group to group. Some of the most universal of these barriers include:

1. languages and communication pattern differences between participants and staff;
2. educational differences;
3. attitudes and values about health care;
4. attitudes of service providers and client populations; and
5. familiarity with health and client populations.

Detailing the many ways in which these issues may affect program design with each possible target population would contain enough information for several books, so a brief illustration of each will be provided. The planner and service provider must rely upon community members to provide accurate and useful information relevant to a particular population.

When considering language and communication pattern differences, it is important to provide materials which are in the primary language of participants or at least to supplement English materials with illustrative or clarifying information in the native language. Training materials must be presented in a culturally compatible way. If extreme deference to elders is appropriate, then

trainers and service providers must incorporate this principle into their own practice to enhance receptivity to their messages. Even with an English-speaking population, communication and language issues may be critical to program success. An older rural population is less likely to be familiar with some of the slang that may be commonly understood by street-wise inner city elderly. Conversely, a lifetime urban resident may be unfamiliar with the language of growing and gardening, canning, and preserving that requires little or no explanation to a farm population. Work with institutionalized elderly who may experience mental impairment requires that the overall training strategy as well as materials and presentations be adapted to the language and communication skill levels of this group.

Educational differences also characterize most of underserved populations. While older persons as a group have less formal education and may therefore experience difficulty with some written materials, underserved groups have a much higher illiteracy rate (National Institute on Aging, August, 1980). Thus, well-developed audiovisual materials are a backbone of training programs. With careful planning, visual materials can leap across literacy, language, and cultural barriers. For example, pictures and samples of spices (and spice mixtures) to be used as substitutes for salt communicates across cultures and takes into account educational constraints as well as providing participants a "hands on" experience.

Many older persons rely upon indigenous health care specialists for primary care as well as health-related advice about prevention. These care providers should be involved in your planning and program activities, if possible. Knowledge of the health beliefs and practices of the target population can help in designing the training program so that it is compatible with community practice.

Service providers and clients may both hold attitudes towards each other that are counterproductive. For example, the service provider who in good faith has made an effort to become familiar with Chicano culture may mistakenly assume that the Chicano elderly are adequately cared for in the extended family. In fact, in many communities, this traditional support system has been stressed to the point of deterioration by geographic mobility, occupational demands, and financial strain. The health promoter should work to strengthen any extended family support that does exist by including family members in health promotion training. Where this is not possible, recognizing and discussing the disparity between the family's expectations, traditional support models and current practice may help to alleviate some of the disappointment and frustration of family members, and also increase receptivity to an organized health promotion program.

Lack of familiarity and experience with the formal health care system may also restrict access to health promotion programs. The location of the program within the health services network may be critical to determining the older person's willingness to participate. If non-medical system care providers are perceived as irrelevant or even threatening to the older person's health and well-being, then locating the program in an outpatient medical clinic may screen out many who would benefit by participation. If older persons in the target group are not currently receiving care from medical specialists then recruiting through health professionals is not likely to secure the desired population. If reimbursement for health promotion services is part of the program design, care must be taken to insure that determination of eligibility, securing physician approval, proof of participation, etc. does not screen out the intended participants who may be intimidated or offended by bureaucratic procedures and structures.

Other issues to consider in developing health promotion programs with these groups are broad organizational and community concerns about the nature of the program structure. Some basic questions to raise and consider include:

1. Should services be provided separately for target population members from the larger community of older persons or should the special needs of target groups be incorporated into integrated programs?
2. How should the neighborhood or target community be defined—geographically or culturally?
3. Is proximity of program delivery site and hence visibility and potential credibility of program to target community more important than integrating health promotion strategies and services in a larger centralized service complex? How does available transportation influence this design?
4. Would it be better to bring multicultural considerations and practices into a mainstream health promotion model or should the health promotion model be integrated with prevailing multicultural health beliefs and practices?

These are basic and sometimes volatile issues to raise and consider; there are no easily identifiable answers or resolutions to the problems raised. However, by recognizing the issues and discussing the alternatives with community members and leaders as well as other relevant health and social service professionals, information and feelings may surface which are critical to appropriate program design and successful program implementation.

This discussion begins to outline the challenges presented by developing health promotion programs with distinct population groups. The important point is not the specific barriers or questions identified, but rather the importance of involving the participant population and other community members in the process and respecting the value of their information.

The consideration of the health status characteristics, cultural values and beliefs about health practices of a particular population may be especially useful as you begin to design and develop a program which is appropriate to your group. Within any cultural or ethnic group, there are diverse needs, interests and beliefs, and your needs assessment and planning process should address this diversity.

Identifying some of the more common and broadly applicable health characteristics and values of the target population provides a useful starting point for planning your development campaign. Several examples are included here as guides to trigger your thinking as you begin planning in your community.

FITTING YOUR POPULATION: SAMPLE PLANNING CONSIDERATIONS FOR SPECIFIC POPULATIONS

Institutionalized Elders

Health Status:

—Two-thirds of institutionalized elderly are afflicted by two or more chronic conditions.
—Major impairments include heart disease, stroke, speech disorders associated with stroke, arthritis, diabetes, and mental disorders.
—Studies indicate the "severely disabled" comprise 5-20% and those "needing some help" range from 50-55% of all residents.
—Approximately 1 in 4 residents have some degree of bowel and bladder control impairment.

Implications for Health Promotion and Programming:

—Encourage creative exercise and body movement, social skill, and interest building techniques, stress reduction through deep breathing and relaxation exercises, and personal growth through meditation and self-awareness techniques.
—Personalize food distribution and meals, housekeeping procedures, and medication rounds.
—Provide lounges or private areas for health promotion activities and private gatherings.
—Encourage home visits and field trips whenever possible.
—Provide books, magazines, and talking books for the visually impaired.
—Encourage social hours as a part of the activity program to promote social interaction.

Resources:

SAGE (Senior Actualization and Growth Explorations)
1713A Grove Street
Berkeley, CA 94709

SAGE began in 1974 and offers health promotion programs to older people both in institutions and in the community. It makes use of a variety of approaches including progressive relaxation, stress reduction, vitality-flexibility exercises, interpersonal communication, and nutrition counseling. They have books for purchase and a videotape for purchase or rental.

American Hospital Association
840 North Lake Shore Drive
Chicago, IL 60611

Publishes *Coordinated activity programs for the aged: A how-to-do-it manual.*

Asian Elders

Health Status:

—The most prevalent health problems among Japanese elders are esophagus and liver cancers, heart disease, colitis, ulcers, alcoholism, and suicide.
—The most prevalent health problems among Filipino elders are arthritis, pulmonary disorders, chronic bronchial conditions, long-term effects of pesticide poisoning, and chronic and spinal back ailments.
—The suicide rate for Asian elders in the U.S. is three times the national average.

Health Beliefs and Practices:

—Chinese medicine emphasizes prevention as opposed to Western culture's emphasis on crisis intervention. The preventive concepts include philosophy, meditation, nutrition, martial arts (kung fu and tai chi chuan), herbology, acumassage, acupressure, acupuncture, and spiritual healing.
—The principles of Chinese healing practices with some variations, are often shared by other Asian cultures.
—In Asian cultures, there is a strong value on the family or community group, and inviduals seek help from these groups first. Utilization of outside agencies and formal help is viewed as an inability to care for oneself or as an inability of the family to care for its elders.
—Values of modesty, self-reliance, and self-effacement inhibit elders from complaining or from expressing pain or need.

Implications for Health Promotion:

—Situate the project or service within the community for convenience and accessibility.
—Involve individual clients, families, and community organizations in all stages of the project.
—Provide bilingual speakers and written materials, and provisions for those who don't read.
—Practice culturally appropriate behavior and demonstrate awareness of norms regarding treatment of elders (e.g., polite form of address, not using first names, serving elders first at meals, etc.).
—Try to integrate Eastern medical practices with Western practices.
—Focus on group interactions and support rather than individual betterment while emphasizing how the concepts and practices being taught will increase self-reliance and independence.
—Deal with basic needs first; set up priorities according to what is needed and be open to helping with problems other than health.

Resources

Asian-American Mental Health Research Center
Ad Hoc Task Force on Aging
1640 W. Roosevelt Rd.
Chicago, IL 60608

National Pacific/Asian Resource Center on Aging
927 15th St., N.W., Room 812
Washington, D.C. 20005

Black Elders

Health Status:

—Leading causes of death in elderly Blacks are hypertension, heart disease, and diabetes.
—Nearly three-fourths of the deaths in 1977 among Black females 65-79 years of age were due to diseases of the heart, malignant neoplasms, and cerebrovascular diseases, as were nearly four-fifths of deaths among women 80 years and older.

Implications for Health Promotion Programs:

—Actively involve individual clients, family, church, and other community organizations in all stages of a health promotion project.
—Use natural networks that already exist in the community.
—Provide services and classes at times and places which are safe. Develop and incorporate strategies for helping elderly feel safer about leaving their homes. Provide transportation when possible.
—When planning nutrition programs for Black elders, be aware that many Black elders live and eat alone.
—Promote blood pressure checks. Stress and poor diet contribute to hypertension in lower socioeconomic Black elders.

Resources:

National Center on Black Aged
1424 K St., N.W., Suite 500
Washington, D.C. 20005
(202) 637-8400

National Council on Black Aging
Box 8522
Durham, NC 27707

National Urban League
The Equal Opportunity Building
500 East 62nd Street
New York, NY 10021

Native American Elders

Health Status:

—Many deaths among Native American elders result from poverty-related disease and life-long deprivation of adequate nutrition and health-care.
—The leading cause of death for Native American adults are accidents (drowning, fire, hunting, and automobile), diseases of the heart, malignant neoplasms, and cirrhosis of the liver. The incidence of cirrhosis has increased 221% since 1955.

—Certain diseases occur more frequently in specific tribes: osteoarthritis among Blackfoot and Pima; diabetes among Papago; gallbladder disease, trachoma, and dislocated hips among Southwest Indians.

Health Beliefs and Practices:

—In the Indian culture, "medicines" are closely associated with religion. The Shaman (medicine man) considers the health of the whole individual, including the body, the mind, and the ecological sphere.
—The treatment process, which may include religious rituals, is guided toward bringing the "patient" back into balance with his or her surroundings.
—Most tribal members believe they must maintain harmony with nature by frequent and regular ceremonies in order to prevent evil from coming to them. Maintaining and restoring harmony with the universe is the basis of complete health.

Implications for Health Promotion:

—Actively involve families and tribal organizations and councils in health promotion programs.
—Provide bilingual materials and service providers as well as oral and audio-visual presentations for nonreaders.
—Make programs available to elders where they live.
—Involve and support indigenous health providers and their practices.
—Meals are valued times in that they provide companionship and grandparent-child interaction. Nutrition education programs could be designed to support this value for example in a daycare setting.
—Exercise programs may be incorporated into daily life with traditional games and dances, tournaments, hiking and camping trips, and gardening.

Resources:

National Indian Council on Aging
P.O. Box 2088
Albuquerque, NM 87103

White Cloud Center
University of Oregon Health Sciences Center
840 S.W. Gaines Rd
Portland, OR 97201

Latino Elders

Health Status:

—High incidence of poverty contributes to the poor health status of Latino elders.
—Many health problems of this group are the result of heavy farm and manual labor.
—Specific health problems are tuberculosis, malnutrition, and respiratory and back ailments.

Health Beliefs and Practices:

—Traditional Mexican health belief systems embody belief in supernatural powers, reliance on ritual, and knowledge regarding medicinal use of plants.
—Factors which partially determine use or adherence to indigenous medical practices include age, accessibility to health care, degree of acculturation, and language, religion, family generation, and family practices.
—Illness, physical or mental, is viewed as a loss of equilibrium between man and nature, or a loss of balance between the natural and supernatural worlds.
—To maintain these proper balances, Latinos rely upon prayer, relics, and "good faith" as well as herbs and spices to treat illness.
—To maintain good health, the curandero (healer) encourages the consumption of foods that are compatible and which maintain balance in the body, i.e., cool liquids to combat fever.

Implications for Health Promotion:

—Seek to coordinate families, indigenous healers, and natural networks in health promotion programs.
—Use Spanish-speaking educators and materials written in Spanish, whenever possible.

—Recognize that the Latino elder (as do many elders) may possess a strong desire for independence and may not want to accept free care without having some way to reciprocate.

Resources:

Asociaíon Nacional Pro Personas Mayores
1730 West Olympic Boulevard, Suite 401
Los Angeles, CA 21064
(213) 487-1922

National Association for Spanish-Speaking Elderly
1801 K Street, N.W., Suite 1021
Washington, D.C. 20006
(202) 466-3595

National Coalition of Hispanic Mental Health
and Human Services Organizations
1015 15th Street, N.W., Suite #402
Washington, D.C. 20005

National Hispanic Council on Aging
Sociology Department
Weber State College
Ogden, UT 84408

These samples identify broad generalizations about the populations considered and they may not apply or may apply only in part to your particular group. The best source of accurate information is the potential participant. Generalizations do suggest questions to ask and concepts that should be explored. The resources identified and references listed in this section, similarly, are intended as places to start your search for information and assistance.

DILEMMAS AND OTHER MISTAKES THAT ARE EASY TO MAKE AND HARD TO AVOID

Either you have read the program development guidelines and have tried to follow them or you have had your own ideas and experiences with program development and have used those to direct your efforts. Despite heroic efforts to be reasonable, sensible, realistic and responsive in program development activities, several potential pitfalls and challenges may be lurking in the shadows. The Wallingford Wellness Project experiences and those of other health promotion program providers who have contributed to the knowledge base of this model suggest some common mistakes and dilemmas that are noteworthy to consider. Some of the most challenging are discussed in this section.

Not Enough Planning Time

The pressure is on to get a program going, right now! Several sponsors have pledged support of various sorts, the activity director of the senior center is anxious to get started and needs to schedule the facility for the next six months, and a number of potential participants have expressed interest and have helped to draw up an outline of the proposed program. All of these systems are saying . . . GO!!!

The senior center advisory board still has some questions about the validity of having a volunteer teach nutrition. Should a registered dietitian teach those classes? There isn't anyone on board yet to facilitate the Personal and Community Self-Help component, and the doctor on your advisory board isn't sure what assertiveness training has to do with health anyway. A local family practice group is willing to subsidize the materials cost of any of their patients who participate. What about charging fees to some people and not others? All of these systems are saying . . . WHOA, hold on just a minute!

The tradeoffs between going full steam ahead and taking the time to address some tricky questions are not always clear. Our experience tells us that if you are in doubt about how fast to proceed . . . slow down. And if you are not in doubt, then stop. Look at the whole picture carefully and then proceed with caution. Any potential problems that arise at this stage are likely to be comparable to a low grade fever. If you do not find out what the source is and take preventive action, you may end up with a full blown program illness that requires intensive treatment.

If you absolutely must get something started before you are really ready, try to scale things down and actively involve the people who are planning with you in this prioritization process. It is much easier to build on a small success than to recover from a large scale disaster. In the process of achieving a manageable size success, you will pick up new advocates and supporters that will help you build the comprehensive program you want to implement.

It is sometimes difficult to sell your sponsors on the idea of an extended planning period. Be as clear and as concrete as possible about just what you intend to do with that time. Try to put the need for planning into a relevant frame of reference. A company that sells a product will understand the need for a period of product development, product refinement, and a marketing analysis, not to mention a consumer survey. You might remind them that despite clear need for a product, and an accessible consumer population, if the consumers do not like the way the product looks, tastes, smells, it will not sell.

The need for allowing adequate planning time does not cease with the beginning of your program or project. Time must be built in to address participant feedback, staff evaluations, and the suggestions and concerns of other community members. Securing the initial support for a program is a place to start planning, not a place to finish. A market analogy again: a good-looking product may sell once, but defects and possible improvements are also bound to emerge.

It is repeat sales and consumer satisfaction that will ultimately keep the program alive and well. It takes time to incorporate the learning that you obtain as a program evolves. And it takes time to communicate to others the importance of revising and improving and changing the program to better meet the needs of participants and staff. Remember, you, the provider, trainer, or administrator, are not alone in this process. The meaningful involvement of staff, consumers, and sponsors is critical. Involvement provides the broad base of support needed by staff for continuing to give their best to an always challenging program.

Delivering a Prescription Instead of a Participatory Program

The needs of staff, participants, and sponsors almost always outweigh the utility of a rule or regulation for program development and delivery. The multiple and complex job pressures of the planner and service provider can easily result in taking a rigid and prescriptive stance about program structure, content, or method. In the midst of conflicting, ambiguous, and contradictory suggestions from all sides, the easiest response is to settle in and defend your approach for all you are worth. It is an understandable response, but the consequences are often unfortunate. Soon, you may be the only one behind your approach, with others feeling as though their investment in the program has been discounted. This is not to suggest that you blow like a leaf in the wind, changing directions willy nilly. Rather, try to embody the participatory, flexible and responsive model in all aspects of the program.

If the nutrition department of the local college is up in arms because you included the Pritikin diet in your reading list for nutrition and not the USDA *Dietary Guidelines*, try not to: ignore them; dig your heels in and say this is the reading list we developed and it cannot be changed at this point; pay lip service to their input and go right on as before. Instead, invite a representative of the department to speak to the nutrition class about the controversial nature of nutrition research. Invite their review of your educational materials and solicit their suggestions for how you can present a balanced view of the issue that still allows participants to understand the diverse information available and exercise their own informed judgment about what material to read. Discuss the issue with participants in the nutrition class and solicit their suggestions for how to deal with the situation.

Another example . . . in your efforts to be flexible, participatory, and non-prescriptive you may plan a totally open and unstructured format for your training sessions. If participants complain about the lack of structure, the lack of organization and their desire for more didactic presentations, pay attention. A participatory educational model is new to many people and the price of force-feeding its principles may be wholesale absenteeism. If participants do not complain about the program structure, ask them for feedback. Remember that soliciting feedback in this model implies that you will pay attention to it and respond whenever possible.

Just as participants may be uncomfortable at first with a new approach to promoting health, you may be uncomfortable with taking a flexible and responsive stance in terms of program design and delivery. You may suggest to participants that they try out this approach, think about it, give it a chance, and criticize it whenever appropriate. Try out those suggestions on yourself and your fellow staff. It is scary to ask for feedback and even scarier at first to attempt changes on the basis of that input. It gets easier with practice. And ultimately it is less scary than finding yourself all alone on a limb defending a position that you are not even really sure is what you would have wanted it to be.

Turf, Investment, and Cultism

Resistance to the idea of community-based health promotion programs exists, just as support for the idea exists. Often this resistance comes from other people who are involved in related activities. Sometimes it is a result of not knowing what your program is really "up to."

Remember, a health promotion program is not necessarily competitive with other health services. A health promotion orientation enhances delivery of health care services. For example, health screening with a nurse practitioner can be highly educational for a client or it can be depersonalizing and mystifying—the quality of service depends on the practitioner's intention and skills and the client's participation. Rather than simply giving test results, the health promotion-minded practitioner shares screening information and skill-building techniques to help the client improve his or her health status. The client's participation in the change process influences the nature of the health care relationship.

Openly sharing information about the program, its intent, and its content will usually lessen opposition. Many fads, and even "health cults," influence the public perception of the health promotion movement. Convey to the community how your program relates to some of the more extreme or "trendy" aspects of wellness. For example, in presenting information about good nutrition, be clear about what you are advocating. Organic foods may be more wholesome, they are also almost always considerably more expensive.

Respect the need and right of participants to exercise choice based upon full information. Jogging and running may be the exercise trend of the moment, but there are many other ways for people to exercise aerobically in ways that do not involve expensive running shoes, featherweight outfits and the bone-pounding of running on city pavement.

If you do incorporate controversial health practices, be aware that these may necessitate more public explanation than emphasizing common sense approaches to positive health behaviors. While some people may be excited by these aspects of your program, you are less likely to recruit the ordinary participant who may find a program which incorporates stress management and communication skills to be risky enough.

Of course you want to avoid as much resistance as possible in setting up your program. But what if it can't be avoided? You have tried persuasion and you have tried to enlist the involvement of those who question your approach so that you can design an acceptable program and you have had to recognize that you have not made any headway with a particular important person or a powerful group. Should you give up? Probably not. Try to understand where the resistance is coming from and why. Share this information with staff, participants, and other influential supporters. Seek their support as you continue planning and developing your program. Develop a long-range plan for cultivating support, keep communication channels open and continue to share information. If active resistance means you have to start small, then do it. Your program can grow with the community's growth in awareness, and your own growth. Remember that you may be making mistakes also; inexperience, lack of information or inadequate resources can be overcome with time. As you expect participants to learn from mistakes, challenges, and frustrations, you can learn too.

The Program Development Guidelines provide you with a place to start. They result from our own experiences, including many mistakes, in developing a health promotion program and from the learning that other service providers and planners have shared with us. As you add to these guidelines, the successes and mistakes of your own efforts, share these with other pioneers in health promotion and with older adults. The more we share our knowledge, the closer we are to providing the most effective, responsive programs possible.

The emphasis on participant-oriented, community-based health promotion is fairly new. Be patient when you can, patient with the community, with other service providers, and with yourself. The goal is a long-term change toward greater health. Model the skills you share with participants: openness, assertiveness, knowledgeability, and nondefensiveness, and this change will happen.

Changing Lifestyle: Learning Health Promotion Content

The preceding section on "Program Development Guidelines" has identified the most important considerations in establishing the basis for a successful health promotion program for older adults. This section will identify the major considerations in *designing* and *implementing* the program with the specific population for whom it is intended. It is assumed that all the essential preparations discussed in the "Program Development Guidelines" have been completed and that the participants have been identified. A program must now be developed and delivered to meet their specific needs.

Ideally, a health promotion program will have a broad approach to lifestyle changes—one which includes classes and activities that address the whole human being, rather than just focusing on one isolated behavior or risk factor. An approach which incorporates, for example, physical fitness, stress management, nutrition, and personal and community self-help will have maximum impact on promoting wellbeing.

Health promotion activities are appropriate for all older individuals, regardless of existing level of health. Although health promotion programs usually emphasize interventions to prevent diseases associated with lifestyle behaviors, they also, especially for older people, include helping those with physical impairments or chronic conditions to maintain and/or improve those conditions.

Minimizing the effects of chronic conditions requires the individual's continued cooperation and involvement in his/her health behavior. The attitudes, beliefs, and values that elders have about health behavior may influence their motivation to attempt and maintain healthy lifestyle changes. A health promotion program should therefore identify and clarify these values, beliefs, and attitudes. Similarly it is important for facilitators to be involved in clarification of their own values. Individuals with negative attitudes and beliefs about aging and limiting expectations about how much elders can or should do to pursue healthy lifestyles will be less effective models and facilitators.

There are many different approaches to delivering health promotion content. Several approaches are discussed in the section, "Variations on a Basic Theme." As the title suggests, these approaches are not mutually exclusive; they represent differences in training activities and techniques. For example, one approach may rely on topic presentation as the primary focus of activity, another approach may emphasize the use of persuasion activities, and another may focus on acquiring self-management skills. The best educational models are frequently well-orchestrated integrations of several distinct approaches (e.g., self-care, peer advocacy, support networks, and community activation) and several different activities (e.g., communicating information, persuasion, acquiring self-management skills, establishing rewarding and dependable consequences, and mobilizing community resources).

The staff and participants of the Wallingford Wellness Project developed an educational approach which we describe as "participative learning." This approach represents a blend of the best of the methods and activities discussed in "Variations on a Basic Theme." Participative learning is found to be effective in facilitating comprehensive lifestyle changes among older adults and in responding to and incorporating participants' diverse interests and goals in health promotion.

A key to understanding the participative learning approach is in viewing the older adult learner not as a passive recipient but as an active collaborator in the learning process. In the Wallingford Wellness Project, older adults were expected to actively participate in the planning, implementation and evaluation of class sessions. For example, one or two participants frequently were asked to examine, with the facilitator, the feedback from the last class session, incorporate that feedback into planning for the next session, and then to help with the next session as co-facilitators. This type of participation is crucial if the training program is to remain flexible and responsive to the needs of an older adult population.

The participative learning approach is based on a laboratory method of learning. Staff using this approach strive to create a low-risk, supportive environment within which older adults, other class members, and staff themselves feel encouraged and "safe" as they try out new behaviors, and develop and choose actions which are most effective in achieving healthier lifestyles.

Learning through experience is emphasized. If experiential learning occurs in situations wherein the learner is exposed to risks in a supportive environment, there will be no fear of failure. For example, if a participant shares with the group an unsuccessful attempt to obtain needed information from a doctor, the group may provide encouragement to the individual and suggestions on how to improve the situation. Using the group's collective wisdom and support for problem-solving may allow the individual participant to risk challenging situations.

Learning from mistakes, what does not work, or what works poorly, can be a successful experience. For example, meditation may not work to reduce stress for an individual, whereas aerobic dance may work very well. Both must be tried to determine which works best. By coming together as a group into a low-risk learning situation, older adults are able to freely share their experiences, discuss common concerns, and compare solutions and alternatives. Thus, the learning process takes a participant-directed focus, and the participants learn from each other as well as from the staff or guest facilitators.

Group participation also provides opportunities for the social interaction that is important to most people. A sense of comradery often develops through shared experience which, in turn, helps to build trusting relationships. The result is a supportive atmosphere in which new ideas and behaviors can be explored and tried out. It is for these reasons particularly that we advocate the participative learning approach.

The health promotion content and the participative learning approach form the two bases for affecting lifestyle changes in older adults. Through discussion, skills practice, and experimentation, health promotion content is learned in a way that can be incorporated into daily life.

Values Underlying Participative Learning

Three aspects of a participative learning approach give the potential user a more complete understanding of the method: values underlying participative learning; designing for participative learning; activities, tools, and techniques for participative learning.

Frequently the values or assumptions which guide an educational approach are not explicitly identified. The resulting lack of clarity gets in the way of the facilitator fully understanding or effectively using the approach. The following list of values or assumptions are intended to provide the potential user with a more adequate understanding of the participative learning approach used in the Wallingford Wellness Project.

The Learning Process:

The learning process is most effective when:

1. There is an agreement between participants and staff to work together as equals and share responsibilities for the success of the course.
2. Participants give and receive feedback with one another as well as with staff.
3. Both participants and staff approach the learning situation with an open mind and a willingness to experiment and experience various activities before drawing conclusions as to their usefulness.
4. Agendas are developed and agreed upon in advance by both participants and staff, and are also flexible enough to be changed based on the needs of the group.
5. It relates to and engages the whole person, i.e., emotions and values as well as the intellect.
6. A variety of methods, activities, and techniques are used to meet the diverse needs of participants.

The Older Adult Learner:

1. The older adult has valuable life experiences, knowledge, and resources which should be identified and used in the learning process. Older adults who have already experimented with lifestyle changes should feel free to share their successes and failures with others in the class.
2. The older adult usually prefers to assume an active role in the learning process, even though some initial resistance may be expressed because of unfamiliarity with the participative learning approach.
3. The degree of participation and involvement of older adults influences their motivation and rate of changing lifestyle behaviors.

The Staff:

1. It is important for staff to demonstrate or model healthy lifestyle behaviors. Much learning occurs through effective role modeling.
2. Modeling by the staff of a desire for and use of the feedback process is an important part of the learning experience.
3. There are important advantages to having a staff which is interdisciplinary, multiracial, intergenerational, and both male and female.
4. An effective staff motivation for teaching older adults about health promotion is the desire to empower participants, to increase their capacity to develop healthy lifestyles and satisfactorily shape the remainder of their lives.

The Learning Environment:

1. Sufficient space should be available so that participants can easily move about and work in pairs or small groups without disturbing each other.
2. The facilities should be as flexible as possible so that various methods and activities can be used. At a minimum, chairs and tables should be movable to allow for variations in how participants interact with one another.
3. The physical environment should be sensitive to special needs of participants (e.g., wheelchair access).

Design for Participative Learning

Designing a health promotion program which incorporates the participative learning approach requires three levels of decisions. First, there are decisions about the overall goals. Second, there are decisions about activities to reach the goals, the sequence of activities, and specifically what each activity involves. Third, there are decisions about specific tools and techniques which would be most appropriate. Examples of potential goals, activities, tools, and techniques can be found in the "Tips for Facilitators" section.

Within these three levels of decisions about design, there are numerous other questions that need to be considered:

1. Who should be involved in the process of designing the program? Should older adults help in the designing?
2. What are the provisions for making revisions?
3. What kinds of options need to be provided? How much individualization of learning activities is needed?
4. How much time is available? How much time is needed?
5. What arrangements can be made for providing followup support?
6. What kind of staffing is needed?
7. What kind of physical environment would best facilitate learning?
8. What kind of activity would provide an effective beginning to the learning process?
9. What kind of learning agreement between staff and participants is most desirable? What responsibilities are defined as "participant responsibilities" and what are "staff responsibilities?" How is the agreement negotiated?
10. What is the best order or arrangement of activities?
11. How much time should be given for each activity?
12. How much time should be given for relaxation?
13. What are the best ways to support the transfer of knowledge and behavioral changes learned in the program to one's entire life?
14. What are the best ways to provide a meaningful end to the learning experience, an ending which is at the same time a beginning to a new, healthier lifestyle?

Activities, Tools and Techniques for Participative Learning

Unlike most teachers, facilitators using the participative learning approach do a minimum of lecturing or instructing. The learning process is a collaborative one in which the facilitators and participants work together to shape the learning experience. Facilitators continually tailor course content to meet the needs and interests of the participants while recognizing differences in age groups, ethnicity, knowledge, values, and life experiences. Facilitators guide, suggest, stimulate action, and serve as resource persons. They avoid providing all the answers by encouraging participants to view each other as resources. Freedom to agree, question or not question, speak or not speak, and guide the group to alternate ways of thinking or acting are all identified as rights of both facilitators and participants.

A major responsibility of facilitators is to provide participants with opportunities to have new experiences and to try out new behaviors in an atmosphere which is supportive, caring, and non-threatening.

Another major responsibility of facilitators is to suggest and select activities, tools, and techniques for participative learning. This responsibility is an integral part of the designing task. Activities are a sequence of steps chosen to accomplish a specific learning objective. A complete learning design consists of a sequence of activities. "Methods" and "exercises" are terms which are frequently used interchangeably with activities.

One common activity is lecturing. Other activities include: small and large group discussions; panel presentations and discussions; brainstorming; audio-visual presentations; demonstrations; role plays; and interviews. Please refer to the Tips for Facilitators for detailed descriptions and uses of these activities.

It takes considerable knowledge, skill, and experience with the participative learning approach in order to most effectively employ it with older adults. A strong value commitment to the approach is also important. Since many of us have probably acquired our health knowledge through traditionally overused means (lectures and reading), we may have a tendency to communicate our knowledge through similar means. If you have limited experience with this approach, we urge you to seek the assistance of those who have had more experience. They can act as co-facilitators or consultants or simply as a source of encouragement and moral support.

The Pillars of Health Promotion

As you progress through *A Healthy Old Age* you have discovered the why, when, where, and how of health promotion programming and implementation. *What* should a program include? As mentioned previously, health promotion programs for older adults can encompass everything from a simple blood pressure screening clinic to a multi-level community-based approach to wellness. Based upon our experience at the Wallingford Wellness Project, four themes have been identified as the pillars (the basic necessities) of health promotion; stress management, nutrition, physical fitness, and personal and community self-help.

In this section, the interrelationship of these themes, some common health concerns of older adults and personal wellbeing are examined. This information will, hopefully, set the stage for planning your own program. Further information and resources for the pillars of health promotion and some common health concerns are located in the Resources section of this manual.

Stress Management

For many, old age is accompanied by a loss of financial security and status, increased problems with health and disrupted social supports. These circumstances, coupled with negative stereotypes of aging and resulting feelings of "having outlived usefulness" can contribute to high levels of stress and tension. For some older adults, these stresses are compounded by continuing cultural discrimination and the conflict between traditional and contemporary social values and practices.

Stress is a significant risk factor for heart attack and stroke and a contributor to other chronic illnesses. Consequently, reducing or managing unnecessary or excessive stress is desirable for improving health and for preventing or managing chronic diseases.

Learning how to identify stress and managing unwanted stress by practicing and applying relaxation techniques in daily life would be useful outcomes for a health promotion program that includes stress management as one of its components.

Nutrition

A good diet is a cornerstone of good health. Studies show that poor dietary choices, such as eating too much fat, salt, and sugar are related to obesity, tooth decay, diabetes, heart disease, and other illnesses. In a more positive light, eating healthy food provides a body with energy, strength, and an increased capacity to resist infection and disease.

Today's most prevalent nutritional problems, for people of all ages, are over-eating and ill-advised food choices that provide calories but inadequate nutrients. Elders may also have to cope with a variety of physical and social circumstances that affect proper nutrition. Loss of natural teeth, reduced ability to digest food easily, decreased capacity to see, smell and taste, physical limitations that hamper preparing food and limited food budgets are just a few examples of problems that may get in the way of insuring a good diet for an older person.

A nutrition education program designed specifically for the elderly should take into account the special dietary and physical needs of its participants as well as provide information and resources that support a change to healthy food habits. The cultural preferences of participants for particular foods and food combinations will also influence the content and orientation of the nutrition program. For example, recognizing the reliance upon rice and fresh vegetables as staples for Asian elderly guides the development of presentations on complementary proteins and meat substitution.

Possible goals for a nutrition course in a health promotion program include: (1) gaining knowledge of good dietary guidelines; (2) providing useful information with which to make educated choices regarding diet; and (3) making positive changes in dietary habits through a gradual and developmental process of reducing, substituting and eliminating less healthful foods for healthier foods.

Physical Fitness

According to the President's Council on Physical Fitness and Sports, a high proportion of the physical decline attributed to aging is actually the result of inactivity and inadequate nutrition. A regular program to improve heart and lung endurance, muscle strength, and flexibility can halt or even reverse that physical decline (Keeler, 1980). The American Heart Association recently added inactivity as a risk factor for coronary heart disease. Regular exercise may contribute to reducing such other risk factors as stress, obesity, and high blood pressure.

Inactivity is accompanied by a cycle of poor circulation, muscle atrophy, reduced strength and endurance, reduction in density of bones, and fatigue. Regular exercise interrupts this cycle, affording a person greater strength, more agility, and improved balance which can be used to stop falls or to mitigate injuries. Through physical activity, a person increases the likelihood of remaining mobile and of retaining an independent lifestyle in old age.

Education and instruction are essential to reach the majority of older people. Many people do not know that constitutes a good exercise program or the amount of exertion necessary to obtain desired physical changes. Only 30% of people age 50 and over are involved in regular exercise. Unfortunately, 68% believe they get all the exercise they need (Harris, 1978). Therefore, some goals for a health promotion program's physical fitness component should include: (1) gaining a basic understanding of the physiological effects of exercise; (2) increasing flexibility and strength of muscles, joints, and bones through learning and practicing various stretching exercises and; (3) increasing heart and lung fitness by regularly engaging in aerobic exercises such as brisk walking, swimming, and biking.

Personal and Community Self-Help

Self-responsibility plays a key role in reducing health risks and promoting health. According to former Health, Education and Welfare secretary, Joseph Califano, "You the individual can do more for your own health and wellbeing than any doctor, any hospital, any drug, and any exotic medical device" (*Healthy People*, 1979, p. viii). Personal and Community Self-Help looks at the issue of self-responsibility as it relates to the health of the individual *and* the health of the individual's community and environment. The purpose of this component in a health promotion program is to empower program participants so that they may assume more control over both personal health and the health and safety of their environment.

Two distinct yet interrelated parts of Personal and Community Self-Help (PCS) are assertiveness and other communication skills and the application of these skills to community and social issues. The PCS course teaches participants skills in constructively asserting their wants and needs. This ingredient is indispensable if participants are to translate the useful knowledge gained in stress management, nutrition, and exercise into lifestyle changes. Knowledge that a particular behavior or action could contribute to one's health is useful only if one can assert his/her right to act accordingly. For example, an older woman decides to reduce her intake of high-fat cheeses and substitute those with skim-milk cheeses. However, the only grocery store accessible to her does not carry skim-milk cheese. In the PCS class, she can learn and practice the assertion skills needed to go to the store manager with the request that skim-milk cheeses be stocked.

Interpersonal communication skills can be used to address social issues of concern to participants. This is important because, in spite of significant health-enhancing changes made in personal life, one is still subject to "unhealthy" environmental conditions. The PCS component does not tell participants which environmental conditions are unhealthy. It can, however, help participants clarify their values, gain information about issues of concern to them and learn techniques for addressing these issues.

Enabling people to assume more control over their lives begins by helping them perceive themselves as effective "change agents" having the right and ability to influence internal and external forces affecting their lives. Thus, learning how to assert feelings and values without hurting others and learning how to ensure the health and safety of both the individual and environment are goals of the Personal and Community Self-Help component.

Addressing Common Health Concerns

Although the pillars of health promotion form the primary content of a comprehensive wellness program for older adults, the program can and should address those health needs, concerns, and problems common among the elderly. Inclusion of a session or a series of sessions on the following health concerns will enhance your programming and make it relevant to your target population.

Accidental Injuries
Alcohol Abuse
Arthritis
Cancer
Diabetes
Insomnia
Medicine Management
Normal Changes of Aging
Oral Health
Sensory Loss

Foot Care
Heart Conditions
Hypertension
Tobacco Smoking
Weight Control

Fact sheets containing information and resources for delivering a session on these topics are provided in the Resources section under "Common Health Concerns."

It is important to see both the pillars of health promotion and the common health concerns as integrated parts of one program. The task of balancing and integrating the different content areas requires an understanding of the participants' needs and interest and their active involvement in the planning process. For example, diabetes and hypertension are particularly prevalent among Black elderly. Sessions which focus on these topics may be of special interest to participants.

The common health concerns can be incorporated into your health promotion program by tying a specific topic to a core content subject. For example, in the nutrition class you may want to expand upon the materials presented in the four nutrition class outlines by offering one or more presentations on diabetes, oral health and/or weight control. Likewise, a stress management course could be enhanced by additional information on alcohol abuse, hypertension and/or insomnia. The common health concerns, therefore, are offered as complement to the pillars of health promotion. They are tools to be used to expand, enhance and further integrate your program offerings.

Synergy: Putting It All Together

A comprehensive, integrated program that incorporates all four pillars of health promotion and common health concerns is recommended. A combined program of health education and group and environmental support for lifestyle change in each of these areas will contribute greatly to enhanced wellbeing and minimized impact of disease and disability.

At first glance, the task of developing a health promotion program that encompasses all four pillars may seem too big to tackle. Limited staff, financial and support service resources may hamper development of a comprehensive program in your setting. However, developing a program which includes all four areas will ultimately take advantage of a *synergistic* effect in promoting health. In other words, the whole will be greater than the sum of its parts.

To illustrate the concept of synergy as it applies to health and wellbeing, let us examine how combinations of unhealthy habits affect us. The three leading causes of death in the United States are heart disease, cancer, and stroke. These are among the array of "lifestyle" diseases that are caused in part by several risk factors including smoking, overweight, lack of exercise, a diet high in salt, fats, and sugar, high levels of stress and tension and high blood pressure.

The deleterious effects of these unhealthy behaviors multiply with the number of risk factors present. That is to say, the more "sickness-styles" a person has—such as poor diet, high stress, and lack of exercise—the more risks that person faces in developing a chronic illness, much more so than if the person had only one risk factor.

Poor diet, inadequate exercise, high levels of stress and tension and environmental hazards are common risk factors; most people need to make healthy changes in at least one of these areas if not more. Conversely, health-promoting behavior changes practiced in tandem are cumulative in effect: The healthier your diet is, the more weight you will lose which will increase the physical benefits of exercise, which will, in turn, reduce stress.

A health promotion program which addresses the need for healthy lifestyle changes in the areas of nutrition, physical fitness, stress management and personal and community self-help and which provides the skills-training and support needed for reducing risk factors, will enhance the general level of "wellness" among program participants.

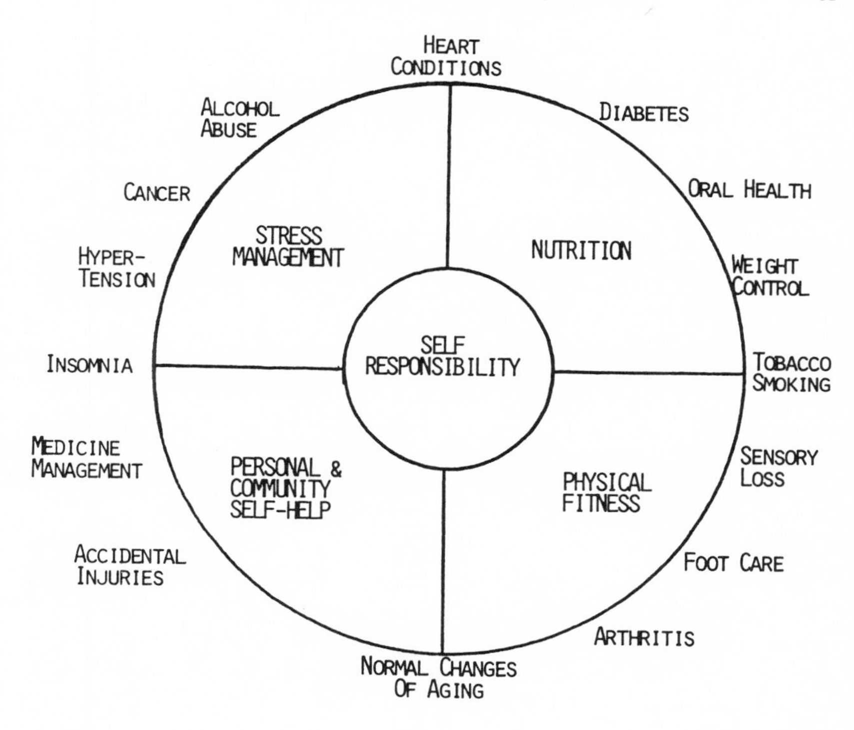

FIGURE 3. The pillars of health promotion and some common health concerns of older adults.

RESOURCES

Variations on a Basic Theme

The basic theme is promoting health and wellness through self-activation and empowerment. There have been several variations on this basic theme with respect to older adults. The purpose of this section is to review self-care, peer advocacy, support networks and community activation as they relate to health promotion.

Below are brief definitions for the four "variations on a basic theme" presented in this section. One basic difference in these approaches is in the number of individuals sharing the responsibility or collaborating in the process of developing healthier lifestyles. For example, self-care emphasizes the role of each individual in lifestyle change. Peer advocacy focuses on older peers as catalysts and social supports in the change process. Advocates of the support network approach to lifestyle change focus on building broad-based support systems for older adults. The community activation approach envisions change at the community or organizational level and involving large numbers of older adults.

> *Self-Care.* A process whereby a lay person can function effectively on his/her own behalf in health promotion and [disease] prevention and in disease treatment at the level of the primary health resource in the health care system. (Levin, 1979)
>
> *Peer Advocacy.* A means of supporting the elderly client's self-worth and dignity so that he/she is able to maintain independent status in today's world. (Bolton and Dignum-Scott, 1979)
>
> *Support Networks.* A set of relations involving the giving and receiving of objects, services, social and emotional supports defined by the giver and receiver as necessary or at least helpful in maintaining a style of life. (Lopata, 1975)
>
> *Community Activation.* An organized group effort to achieve a common goal that is beneficial to all involved.

On the following pages are more detailed descriptions of each of these approaches. Resources, including some of the many programs which exemplify each approach, are also listed. Of course, each of these programs may incorporate aspects of one or more of the other basic approaches to health promotion, and could therefore be considered, for example, both a self-care and a peer-advocacy program.

SELF-CARE

Proponents of the self-care approach believe that personal health and wellbeing can be significantly improved through action initiated by the individual. Self-care goals are promoting health as well as preventing, detecting, and treating disease. Self-care education draws from and builds upon the knowledge and skills one already possesses; the emphasis is on the individual as decision-maker. As Norris (1979) explains, "the essence of the self-care movement is control, responsibility, freedom, expanded options, and an improved quality of life."

Butler (1979) proposes that self-care be thought of as a series of concentric circles. The inner circle would include those tasks which are part of our daily routine: basic hygiene and dental care; nutrition; simple first aid. For example, substituting low-fat foods for those with a high fat content into the weekly menu plan represents this level of activity. The second ring would include patient education and consumer health education programs which may result in new health enhancing behaviors, e.g., smoking cessation. The third and outermost ring would include those activities which require skills traditionally available only to health-care professionals, e.g., learning to give an insulin injection, taking a blood pressure reading or doing a complete physical exam. These skills can be acquired through educational programs often conducted by a nurse or physician.

While self-care may not cure the ailments that often accompany aging, it may slow the advance of some disease processes and improve the quality of life. When older adults are able to attain maximum control over their bodies, they also gain a sense of greater control over their entire lives, their sense of wellbeing and purposefulness. Self-care skills can help them gain renewed confidence in their abilities. Another benefit of the self-care approach may be a reduction in health care costs resulting from a decreased and more appropriate use of health care services.

Four primary sources of self-care information are available: parents, the media, self-help groups, and professionals. Mothers have probably been the most effective practitioners and teachers of self-care. A majority of our health care practices are learned at home, and much of this learning seems to stay with us for life. The media, especially in recent years, is another source of self-care information. Broad-scale campaigns have been waged to enhance awareness of the health risks associated with lifestyle behaviors and to recommend actions individuals can take to reduce the health risks (DSHS, 1980; Farquhar, 1977).

Self-help groups offer participants a support base for changing behavior in addition to providing information. Williamsen and Danaber (1978) outline four kinds of self-help groups which include aspects of self-care. Patient education groups focus on disease (e.g., diabetes, arthritis, or hypertension) and developing the necessary skills required for management of that disease. Rehabilitation groups help patients with various chronic conditions or disabilities through sharing experiences, support, and methods of coping with the situation. Behavioral modification groups such as Weight Watchers and smoking cessation groups focus on changing all types of behaviors which contribute to poor health. A fourth type of group focuses on reducing risks to disease and includes areas such as nutrition, stress management, aerobic exercise, and general consumer health education. These groups exemplify a self-care approach to health promotion.

While most self-help groups are run by lay persons, prevention and patient education self-help groups are often organized and conducted by health educators and/or other health care professionals. Physicians and nurses conduct a significant amount of education for self-care in clinics, hospitals, and other health care settings.

Self-Care and Self-Help Programs

Self-Care for Senior Citizens
Department of Community and Family Medicine
Dartmouth Medical School
Hanover, NH 03755

This program consists of 13 two-hour sessions, monthly newsletters, and 6 month reunions. Session topics include: responsibility for health care, facts and fallacies of medications, emergencies, common illnesses, chronic conditions, lifestyle change, exercise, nutrition, emotional wellbeing, and discovering and using social services.

Growing Younger
Healthwise, Inc.
P.O. Box 1989
Boise, ID 83702

This program works with neighborhood groups of older adults. A series of group meetings covers topics vital to wellbeing, such as exercise, relaxation, nutrition and self-care. Healthwise markets a training program and materials to assist other organizations which wish to implement similar programs.

Healthy Lifestyles for Seniors
Meals for Millions Foundation
1800 Olympic Boulevard
Santa Monica, CA 90404
(213) 829-5337

This is a six-month program model with twice weekly classes. It emphasizes self-assessment skills for physical and emotional wellbeing; nutrition; exercise; and stress management. Developing a cooperative and constructive relationship between an inter-disciplinary team of professionals and participants to enhance the relevance and success of individual plans for lifestyle change is a distinctive goal of the program. The project has produced a program development manual for service providers entitled, *Healthy Lifestyles for Seniors: An Interdisciplinary Approach to Healthy Aging*, which is available through the Meals for Millions Foundation.

The Arthritis Self-Management Program
701 Welch Road, Suite 2208
Palo Alto, CA 94304

The Arthritis Self-Management Program at Stanford University has developed many materials specifically related to arthritis and self-care.

Self-Care Resources

Kosidlak, J. G. Self-help for senior citizens. *Journal of Gerontological Nursing*, 1980, *6*(11), 663-668.

This article describes a self-care/medical self-help course for the elderly, taking the reader through the entire evolutionary process. Contains invaluable insight and information for anyone thinking of starting such a course.

Levin, L. S., Katz, A. H. & Holst, E. *Self-Care: Lay initiatives in health* (2nd Ed.). New York: Prodist, 1979.

A must for anyone interested in self-care, this book reviews the past, present, and future of the self-care movement. Many issues are explored including professional, ethical, and legal implications. It also containes an extensive annotated bibliography on self-care which has been updated for the second edition.

Sehnert, K. W. *How to be your own doctor—Sometimes.* New York: Grosset & Dunlap, 1975.

A comprehensive and practical guide to medical self-help in which the "activated patient" is defined and explored. Self-help guides for intervention in common illnesses and emergencies are provided along with guidelines for self-examination. The book also contains a listing of self-help pamphlets and guides which are available to the public.

U.S. Department of Health & Human Services. *A guide to medical self-care and self-help groups for the elderly* (Rev. ed.). NIH Publication No. 80-1687, 1980.

A 28-page pamphlet which briefly defines and traces the emergence of self-care and self-help, emphasizing their importance to the wellbeing of older adults. Several self-help groups of interest to the elderly are also described.

Clearinghouses

National Self-Help Clearinghouse
Graduate School and University Center/CUNY
33 West 42nd Street, Room 1227
New York, NY 10036

Provides information on various self-help groups across the country. Publishes a highly informative newsletter, "Self-Help Reporter," bimonthly, as well as several publications of interest. These include:

Annotated Bibliography on Self-Help Mutual Aid, 1975-1977, $3.25.

Building Self-Help Groups Among Older Persons: A Training Curriculum to Prepare Organizers, $15.00.

How to Organize a Self-Help Group, $3.00.

Further information, including a complete list of publications, is available upon request.

National Clearinghouse on Aging
SCAN Social Gerontology Resource Center
P.O. Box 231
Silver Spring, MD 20907

Several bibliographies on the following topics are available upon request: family support, peer counseling and self-help, alternative services, day care, family relationships, group housing, institutionalization, organizations, and service planning/management/evaluation. Up to five bibliographies may be sent per request.

PEER ADVOCACY

Peer advocacy focuses on promoting the supports necessary for an older adult to maintain maximum independence. Enhancing one's self-worth and dignity is an integral aspect of this approach. The advocate helps his/her peers by informing them of available services, helping them secure needed services, instilling confidence, and acting as role models so that individuals will eventually become their own advocates. Peer advocates educate, refer, offer support, and provide companionship in a number of settings including senior centers, nursing homes, hospitals, clinics, nutrition program sites, and frequently in an older person's home or apartment.

The primary strength of the peer advocacy approach is in the advocates' ability to empathize and identify with the persons being helped. Shared life experiences and similar ethnic, cultural, social, and economic backgrounds help peer advocates to overcome many barriers that younger professionals may encounter. Peer advocates provide an example of successful aging and illustrate the ability to assume new roles and remain actively involved in community living.

Peer advocates generally receive some formal training from professionals within the agency which sponsors them. The information, skills training, and values clarification are intended to dispel myths and stereotypic thinking about the abilities and needs of the elderly and to focus on the positive aspects of aging. Interviewing and communication skills are especially emphasized. To be effective advocates, older adults must learn to view their own life experiences and those of their peers as valuable in meeting the new life challenges that come with advancing age.

Peer advocates can have many roles in comprehensive health promotion programs. For example, in the Wallingford Wellness Project, many program graduates are now leading the ongoing classes. These "Wellness Advocates," as they named themselves, supplemented their basic health promotion training program with six additional three-hour sessions in group facilitation and leadership skills.

The advocates also can assist individual participants with specific health concerns. For example, an individual who is having difficulty communicating with his/her health care provider can use the skills of the advocate in several ways: to improve his/her own communication skills so that the client/provider relationship is improved; to act as a mediator or a communication facilitator in face-to-face encounters with the provider; or to direct the individual to alternative sources of the needed information or service.

Recruitment and selection of appropriate advocates is as important as the type of training program offered. The best recruitment results will probably be obtained by using a combination of strategies, for example, notices in newsletters and bulletins; posters in social security agencies, service agencies, grocery stores, public transportation; referrals from professionals; and nominations from the elderly themselves. Selection should be based on clearly identified criteria, for example: physical and mental health, warmth, sensitivity and compassion; ability to accept others with differing values and beliefs; and involvement in the community.

Peer Advocacy Programs

Elderly Homebound Program
The Institute of Study for Older Adults
The Division of Continuing Education
New York City Community College of the City

University of New York
450 West 41st Street
New York, NY 10036

This program, designed to bring continuing education courses to elderly shut-ins, trains older adults to act as "learning companions." Each companion attends two-hour classes for ten weeks and then visits an elderly shut-in who is also taking the course, to review and discuss course content. Course materials are based largely on televised programs since most shut-ins have access to TVs. The companions enrich their own learning experiences and gain satisfaction from helping someone else, and the shut-in receives intellectual stimulation, a feeling of accomplishment, and a reduction in feelings of isolation.

Community-Based Comprehensive Care for the Elderly
Institute on Aging
University of Washington
Seattle, WA 98109

This interdisciplinary team project (nurse, social worker, health educator) involves the Seattle Housing Authority, Visiting Nurse Service, and Area Agency on Aging in training peer advocates to improve access to health and social services for the low-income elderly residing in two publicly financed housing projects and surrounding neighborhoods. By working with existing community-based service providers, the team has developed a peer advocacy model for providing a continuum of multidisciplinary, integrated health and social services to the elderly on a long-term basis.

Peer Counseling for Elderly Persons (PEP)
Westside Senior Health and Counseling Center
2125 Arizona Avenue
Santa Monica, CA 90404

The PEP program, sponsored by the Westside Senior Health and Counseling Center, provides peer counseling to older adults who are experiencing anxiety, loneliness, or depression. Counselors are trained to help their peers remain in the community and enjoy an increased sense of independence and self-esteem.

Turner Learning Programs
Turner Geriatric Clinic

University of Michigan Hospital
1010 Wall Street
Ann Arbor, MI 48109

Begun as a complement to the Turner Geriatric Clinic's outpatient services, peer counselors were trained to organize monthly health education workshops on topics of interest to the elderly. They quickly expanded this role in response to the needs expressed by elderly program participants. The peer counselors now provide social services, medical and health information through workshops, and psychological assistance through individual and support groups.

Peer Counselor Training Project
University of Minnesota
3300 University Avenue, Room #290
Minneapolis, MN 55414

This project has developed a model and a process for selecting, training, and supervising older adults as peer counselors. A packet of training materials that can be used by other programs is available.

Volunteer Health Advocates
Health Education Office
NRTA-AARP
1909 K Street, N.W.
Washington, D.C. 20049

Sponsored by the National Retired Teachers Association and American Association of Retired Persons, volunteers establish health education programs for community groups, nursing homes, day care centers, churches, etc.

Senior Companion Program (SCP)
Contact: ACTION Regional Office
or call: (800) 424-8580

Low income persons 60 years or older provide services to adults with physical and/or mental impairments. Services provided include linking clients to appropriate community services and providing companionship and caring with the focus on helping homebound elderly to continue to maintain independent living.

Retired Senior Volunteer Program (RSVP)
Contact: ACTION Regional Office

Persons 60 years and older provide a variety of services to other older persons and community agencies.

Peer Advocacy Resources

Besker, F. & Zarit, S. H. Training older adults as peer counselors. *Educational Gerontology*, 1978, *3*, 241-250.

A program of peer counselor training was conducted to evaluate the acquisition of counseling skills by older volunteers. Modeled on similar programs developed for younger paraprofessionals, the course emphasized the learning of three core therapeutic conditions: empathy, warmth, and genuineness; it also contained information on the aging process to reduce stereotypic thinking. The results indicated that the trained groups changed significantly compared to two of the three core dimensions, empathy and warmth, and had scores on all three scales above the minimum necessary for effective counseling. These findings suggest the potential for older persons to provide supportive, peer-counseling services. (Article Abstract)

Friedman, S. The residential welcoming committee: Institutionalized elderly in volunteer service to their peers. *The Gerontologist*, 1975 *15*(4), 362-367.

This paper describes the experience of a group of relatively impaired, institutionalized elderly in a program of volunteer service to their peers. The focus of discussion is upon the volunteer group itself, its weekly group meetings, and the meaning that this kind of group experience has for its members. The group experiences that are described demonstrate the continued potential for change and the ongoing need for self-determination existing in advanced old age and point to the possibilities for both helping and being helped that are available in this type of volunteer program. (Article Abstract)

Ganikos, M. L. (Ed.). *Counseling the aged: A training syllabus for educators.* Washington, D.C.: American Personnel and Guidance Association, 1979.

This is a very complete and informative manual for anyone interested in establishing an advocacy program. The different aspects of aging, special crisis situations, and the special considerations of minority aged are discussed. An entire section is devoted to the role of peer counselors and their training.

Toseland, R. W. et al. A community outreach program for socially isolated older persons. *Journal of Gerontological Social Work*, 1979, *1*(3).

A pilot outreach program was designed to reduce social isolation and meet the social service needs of elderly living in the community. In order to meet the social service needs of 72 socially isolated older persons, peer counselors used a problem-solving casework approach focused on enhancing self-help skills. Community development efforts and networking techniques were implemented to encourage participants to expand their social interaction networks and develop peer support systems. An evaluation of the pilot program suggests that it was successful in decreasing social isolation and in meeting the social service needs of those who were served. (Article Abstract)

Waters, E., Fink, S., & White, B. Peer group counseling for older people. *Educational Gerontology*, 1976, *1*(2), 157-169.

For a variety of reasons, psychological services are largely unavailable for older people who want help in coping with their interpersonal problems. Group counseling, when it is used, proves to be an effective and efficient method for meeting their needs. It also addresses the reality of the limited numbers of gerontologically trained professionals currently available in the field. A recently attainable and largely untapped resource for working with this population is older people themselves. The Continuum Center of Oakland University selects, trains, and supervises older men and women, who serve as paraprofessional group leaders in a self-exploration program offered in a variety of community centers in the Detroit metropolitan area. The service and training programs, and some of the outcomes for clients and group leaders are briefly described. (Article Abstract)

SUPPORT NETWORKS

This approach emphasizes the importance of having support and encouragement for making individual lifestyle changes. Lopata (1975) defines support systems as a "set of relations involving the giving and receiving of objects, services, social and emotional supports defined by the giver and the receiver as necessary or at least helpful in maintaining a style of life." These support networks help us adapt and cope with changes in our lives.

For the elderly, support networks often are not in place or readily accessible. Older adults are often subject to many losses which negatively influence their wellbeing. Longstanding relationships may be disrupted by the death of a spouse or friend, retirement, impaired health, limited mobility, or relocation from familiar surroundings to senior housing or nursing homes. These multiple losses result in decreased opportunity for interaction with others, and often, forced isolation.

Support networks can provide the elderly with the means for bringing stability and meaning to their lives and helping them maintain autonomy. When supportive ties are missing the coping abilities of the elderly can erode so drastically that they decline into dependency. One study found that the lack or presence of supportive care, primarily from family, was the best predictor of placement in institutionalized care (Brody, Paulschok, & Maciocchi, 1978).

Voluntary and informal helping relationships between older adults and their family, friends, and neighbors are critical in providing support to the elderly (Ehrlich, 1979). The family is the primary helper in this "informal" network. According to Shanas, "Family help, particularly in time of illness, exchange of services, and regular visits are common among old people and their children and relatives whether or not they live under a single roof" (1979). It is the "emotional bond" between children and their parents which is the critical factor in this assistance.

There are several factors, however, which affect a family's ability to care for elderly members. Today's mobile society has resulted in many of the elderly being unable to live close enough to their children for frequent visits. Because of the historical trend toward smaller families, the elderly have fewer children to call on for assistance. Also, with the growing number of women in the work force, it is becoming increasingly difficult for women to continue assuming the major role in the care of their elderly parents (Treas, 1977).

Women have a longer life expectancy than men. They are more likely to become widowed while men usually remain married up to the time of their death. Women are also less likely to remarry than widowers. Accordingly, women have a lesser chance of retaining supportive relationships with spouses than do males (Monk, 1979).

Longer life expectancies for both men and women have created a wider age range among older adults. It is not uncommon to find elderly persons in their sixties and early seventies with parents still living. This may present a major challenge to adult children, who are themselves undergoing the changes of retirement, etc., to continue to provide for their aged parents (Monk, 1979).

Many older persons also rely on friends and neighbors for needed assistance and support. Regardless of who the helper is, natural networks are very important sources of support for the elderly. Persons with many informal supports are found to have a more positive self-concept, be emotionally healthier, more likely to plan for the future, participate in more leisure time activities, and be less likely to feel depressed or alienated. They have a greater feeling of security because they know they have someone to turn to in time of crisis or illness (Stephens, 1978).

Clubs, organizations, church groups, and self-help groups can also be great sources of assistance to older individuals. For those with few primary supports these groups can be very effective in assuming many of the functions of a natural network. A few examples of support groups are: Residential Welcoming Committees for new residents of nursing homes; Neighborhood Watch Programs, where residents keep tabs on their neighbors and work to reduce crime in their neighborhoods; Widow Support Groups to help recently widowed persons adjust to their loss; and the Gray Panthers, who actively lobby for local and national legislation to benefit the elderly and work to improve the self-image of older persons by providing positive age group identification.

Mutual aid and support groups are important ingredients in comprehensive health promotion programs. Whether it is family members, old friends and neighbors, a new acquaintance and fellow classmates who provide encouragement for healthy lifestyle changes, their support can be critical.

This support can take many forms, some carefully structured, others a natural evolution. The family group which is willing to try a healthy new alternative to grandmother's famous marzipan at Sunday dinner is supporting her efforts to make positive changes and the old fishing buddy who does not criticize his friend for substituting juice for the traditional two six-packs of beer is also providing support.

Classmates who form an informal "walking club" to continue their aerobic walking after formal training ends are creating a support network to help each other maintain the healthy changes they have undertaken. The group that negotiates with the senior center director to provide a meeting space for an ongoing health promotion discussion group is creating its own structured internal support system for making and maintaining positive changes. Individual and group support for healthy living styles is necessary for most people, particularly as they attempt to change the habits and behaviors of a lifetime.

When help is needed which is beyond the capabilities of informal networks, formal support systems are called upon. These supports are composed mainly of community service agencies that provide alternatives to premature institutionalization. These services include, for example: emergency food and shelter programs; home-delivered meals and grocery services; adult day care centers; visiting nurse services; homemaker services; congregate dining and lunch programs; transportation services; legal assistance; housing assistance and rent subsidies; telephone reassurance; and friendly visitor programs.

Support Network Programs

Elder Project
Kent School of Social Work
University of Louisville
Louisville, KY 40292

This program enhances and strengthens networks among older persons in a specific neighborhood by involving small groups in weekly educational meetings. Further information is available upon request.

Natural Supports Program
Community Service Society
105 East 22 Street
New York, NY 10010

This program provides support services to relatives and friends who are involved as caregivers to older adults. The caregivers receive education about aging, information about services, benefits to which the elderly may be eligible for, homecare skills, and the opportunity to share their experiences with other caregivers. Information and a list of program publications are available upon request.

SAGE (Senior Actualization and Growth Explorations)
491 65th Street
Oakland, CA 94609

Through the use of many experiential processes, including physical and psychological exercises, participants in SAGE expand their awareness and appreciation of themselves and their sense of community and avail themselves of ongoing health-supporting activities.

Support Network Resources

The clearinghouses listed in the Self-Care section can provide useful information about support networks.

Blythe, B. J. Social support networks in health care and health promotion. In J. K. Whittaker & J. Garbarino (Eds.), *Social support networks: Informal helping in the human services.* Hawthorne, NY: Aldine, 1982.

Describes model projects using formal and informal social networks to promote health behavior. Provides guidelines for developing new networks, summoning existing networks and supporting caregivers.

Caplan, G. & Killilea, M. (Eds.). *Support systems and mutual help: Multidisciplinary explorations.* New York: Grune & Stratton, 1976.

This book provides an excellent review of the mutual help aspect of support systems. Several chapters, including "The Family as Support System," are particularly useful in understanding the importance of supportive relationships for the elderly.

Collins, A. H. & Pancoast, D. L. *Natural helping networks: A strategy for prevention.* Washington, D.C.: National Association of Social Workers, 1976.

In this book, several helping network programs are outlined. The authors provide useful observations and guidelines for working with natural helpers.

Euster, G. L. A system of groups in institutions for the aged. *Social Casework*, 1971, *52*(8) 523-529.

A system of groups for institutionalized elderly is proposed as a means of improving the quality of their lives. An overview of the problems of institutionalization is given along with detailed outlines of the various groups proposed.

Moulthrop, H. E., & Roxborough, J. Network support for the aged: The viable alternative to institutionalization. *Journal of Gerontological Nursing*, 1978, *4*(6), 64-66.

The authors advocate the exploration and development of networks for the elderly as alternatives to institutionalization. Directed at medical personnel, it offers examples of how these networks can be utilized to improve the health and wellbeing of elderly individuals.

Pilsuk, M. & Minkler, M. Supportive networks: Life ties for the elderly. *Journal of Social Issues*, 1980, *36*(2), 95-116.

Recent studies suggest that the increased vulnerability of the older person to physical and/or mental breakdown is related to loss or deficiency in the pattern of supportive ties.

Various programs have addressed the differences in social support for older people in markedly different ways. Six different programs are examined to illustrate the importance of attention to health status, ethnicity, and style of life in the provision of social support. By evaluating the offerings of these programs against the concepts of network theory we are able to show how the needs for social support among the elderly are highly differentiated and deserving of equally differentiated forms of response. (Article Abstract)

Ruffini, J. L. & Todd, H. F. A network model for leadership development among the elderly. *The Gerontologist*, 1979, *19*(2), 158-162.

A network model for the organization of the elderly for self-help and the development of leadership is described. While more research is needed to determine which variables are associated with degrees of willingness to participate in organizations and to help needy peers, the relatively flexible network model described is recommended as the most appropriate for those concerned with organizing *all* types of elderly, even if only minimally, on a large scale, for a sustained period of time. (Article Abstract)

Pamphlets

Irwin, T. *After 65: Resources for self-reliance.* Public Affairs Pamphlet No. 501.

This pamphlet considers the wide range of social services that help the elderly help themselves. As the pamphlet states, "The key to keeping up our spirits and wellbeing as we grow older lies in knowing what resources and services are available, and in using these facilities to advantage."

Ogg, E. *Partners in coping: Groups for self and mutual help.* Public Affairs Pamphlet No. 559.

This pamphlet gives a general overview of the self-help movement, describing several groups now in existence. It provides useful hints on how to establish a self-help group.

These pamphlets are available from:

Public Affairs Pamphlets
381 Park Avenue South
New York, NY 10016

COMMUNITY ACTIVATION

In previous sections, self-responsibility has been identified as essential to improving one's health and wellbeing. There are, however, many factors affecting one's ability to lead a healthy life which are beyond immediate individual control. For the elderly, fixed incomes, housing problems, and high crime rates are examples of a few of the problems they encounter. The elderly, through community activation, can begin to take charge of situations which previously may have left them feeling helpless.

Community activation is self-care, peer-advocacy, and support networks on a large-group basis. It involves an organized group effort to achieve a common goal that is beneficial to all involved. In this context, a community can be a neighborhood, a senior housing complex, a nursing home, church group, or any other group of persons who share a common problem, situation, or interest.

The purpose of a community activation approach is to increase group effectiveness as decision-makers. The approach requires active and responsible involvement in a process of identifying unmet needs, analyzing the availability of resources to meet those needs, and generating proposals for allocating whatever local and/or outside support is needed to alleviate the identified problems. Although knowledge of the existing resources in the community is an essential basis for the community activation approach, frequently resources or services are linked together to create an effect on the identified need or problem.

The Wallingford Wellness Project provides an example of how a health promotion program involves community activation. Through the Personal and Community Self-Help component of the training, project participants begin to examine issues which affect their health and safety. They learn to recognize conditions that may be harmful not only in the larger community environment, but in their own homes and neighborhoods. Members begin by making personal changes such as reading food labels to check fo additives, by recycling, or by using alternate forms of transportation. Through assertion training, participants learn how to assert themselves personally and take charge of a situation. As their confidence and independence grows, they move on to discuss the broader issues such as the high cost of food, ecology, social security, and legislation affecting their community. By planning and prioritizing strategies together, participants begin to explore ways in which they, as a group, can deal with these problems.

The Growing Younger Program in Boise, Idaho also represents a community activation approach to promoting health and wellbeing in an older population. Senior citizens are organized into small neighborhood groups to offer each other long-term support for the good health of the members and the communities. The focus of the groups is on positive health and wellness. Each group has the opportunity to explore exercise, relaxation, nutrition, self-care and other topics vital to good health.

There are numerous examples of community activation projects that can be specifically designed for an older population. Recycling paper, glass, and aluminum not only helps to reduce waste and improve the ecology, but also raises a little money. If transportation of bulky items is not a problem, any group can recycle.

"Neighborhood Watch" programs are another good example of organizing for a community's wellbeing. Community residents watch each other's homes when residents are away, reporting suspicious activities, tagging valuables with identification numbers, and accompanying each other on shopping trips.

Community health fairs are ideal for spreading the word about health promotion and making informed choices about factors affecting health. Sponsoring a fair involves a great deal of planning and coordination with other community organizations. Less experienced groups may want to first get involved with sponsoring a booth at a carnival or street fair.

Community Activation Resources

O. M. Collective. *The Organizer's Manual.* New York: Bantam Books, 1971.

Written as a result of the student movement of the 1960s and early 1970s, this book is nevertheless invaluable for anyone involved in community activation. This book is a guide for organizing and mobilizing communities.

Ruffini, J. L., & Todd, H. F., Jr. A network model for leadership among the elderly. *The Gerontologist*, 1979, *19*(2), 158-162.

The authors describe a network model designed to organize the elderly for community leadership and self-help. The outlined program provides an excellent example of elderly volunteers assuming different levels of involvement and responsibility within the community.

U.S. Department of Health & Human Services. *Toward a healthy community* (*A model fair*), DHHS Pub. No. (PHS) 80-50112, Washington, D.C.: Government Printing Office, 1980.

This pamphlet describes a 1979 health fair sponsored by HEW in Washington, D.C. Included are a discussion of the planning and development stages and an evaluation of the project. It offers excellent tips for any group contemplating their own health fair.

National Council of Senior Citizens
1511 K Street, N.W., Suite 540
Washington, D.C. 20005

Publishes *Criminal Justice and the Elderly*, a newsletter describing crime prevention efforts by older adults across the country.

National Self-Help Clearinghouse
Graduate School and University Center/CUNY
33 West 42nd Street, Room 1227
New York, NY 10036

Publishes *Self-Help Reporter*, a newsletter which updates the activities of groups across the country including barter networks, home maintenance groups, tenant groups, food co-ops, conservation groups, etc.

Self-care, peer advocacy, support networks and community activation are closely interrelated. Community activation strategies, of necessity, incorporate aspects of peer advocacy and support networks. Self-care experiences may provide the motivation to address community needs, such as a medical emergency information hotline, or a community policy of pharmacists willingly providing consumer information on drug interactions.

The approach which is emphasized in a particular program depends upon participant needs, interests, and resources. A health promotion program in a nursing home for the severely impaired may choose to emphasize a peer advocacy approach, with volunteers working closely with family and other health care providers to create the healthiest possible living situation for residents. In this setting, self-care is another viable program component. Even severely impaired residents may be able to effectively monitor some of their own needs and functions, as well as possibly those of roommates or fellow residents. This can be built into support among the residents for the healthy changes which are the goal of the health promotion program.

Like the pillars of health promotion, self-care, peer advocacy, support networks and community activation work synergistically. Incorporating and integrating two or more approaches can increase the impact and effectiveness of your program.

Many other organizations and agencies throughout the country are developing or implementing health promotion programs with older people. These range from hypertension screening and treatment programs to comprehensive multicomponent lifestyle change projects. A few additional programs are listed below which may be good sources of information, program materials, or moral support.

The National Urban League (500 East 62nd Street, New York, NY 10021; (212) 310-9129), in cooperation with Upjohn and the

American Red Cross, is training older volunteers in several cities for hypertension screening and other prevention activities. This project is called the Senior in Community Service Program.

The New York City Department for the Aging through funding from the State Department of Health is sponsoring "The Health Education and Monitoring Program." Information can be obtained from Joyce LaPenn, NYC Dept. for the Aging, 280 Broadway, New York, NY 10007 (212) 577-0859.

Catholic Health Corporation with corporate headquarters in Omaha, NE, administers more than a dozen hospitals and nursing homes throughout the Midwest. "Wellness and Health Promotion" is the number one programming priority at the current time and health promotion projects involving older persons are prominently featured in member hospitals. CHC can be contacted at 1801 S. 72nd Street, Omaha, NE 68114 (402) 393-7661, attn. Susan Cochran.

Seniors' Health Program
Augustana Hospital
Chicago, Illinois
attn: Robert Skeist, R.N.

Augustana Hospital sponsors a health promotion program for older people which focuses on exercise, massage, meditation, and Tai Chi Ch' nan as well as including specific topics such as Medication Management, foot care, and home safety. The program manager, Robert Skeist, has published *To Your Good Health! A Practical Guide for Older Americans Their Families and Friends*, Chicago Review Press, 1980.

Tips for Facilitators

Immediately following this section are four curriculum modules for each of the subject areas we believe should comprise the foundation of a health promotion program: Stress Management, Nutrition, Physical Fitness and Personal and Community Self-Help.

Each of the modules follows the same format for topic presentation:

Session One:	An Overview of the Topic
Session Two:	Basic Strategies for Incorporation into Lifestyle
Session Three:	Specific Technique
Session Four:	Specific Technique

For example, the Stress Management module contains:

Session One:	An Overview of Stress and Stress Management
Session Two:	Basic Strategies to Identify and Manage Stress in Your Life
Session Three:	Relaxation Technique
Session Four:	Time Management

The outlines provide, in essence, a framework of possibilities. Each session contains an agenda, activities, information, group exercises, handouts (which you may reproduce for distribution to participants) and other materials.

Depending upon your resources and your program participants' interests and capabilities, these sixteen sessions may comprise your entire wellness program or they may serve as a starting point for a more extensive program.

The purpose of the "Tips for Facilitators" section is to help course facilitators use the participative learning approach in conducting these sessions. The guidelines and principles presented are based upon the experience of what worked and did not work in the

Wallingford Wellness Project and from readings in participative learning.

This section considers the three most important responsibilities of the facilitator. First, the facilitator must establish learning goals and objectives to promote health among participants. Then, after decisions about learning designs have been made and learning activities and tools have been selected, the facilitator must direct attention to creating the most supportive and nonthreatening atmosphere possible for the participants. Finally, the facilitator must effectively guide interactions between participants so that the learning objectives will be achieved.

The last part of this section describes educational activities and tools which have been successfully used by facilitators in the Wallingford Wellness Project. As defined elsewhere in this sourcebook, activities refer to the steps needed to attain an objective and tools are those forms which participants complete.

ESTABLISHING LEARNING GOALS AND OBJECTIVES

The broad learning goals for promoting health should be formulated even before recruiting participants. Goals you may want to establish for non-institutionalized older adults may include increasing independence, self-worth, knowledge about normal changes that come with advancing age, knowledge of coping skills, and use of problem solving skills in making new adjustments. Goals to consider for institutionalized older adults may be to stimulate verbalization, to decrease isolation by increasing interaction among group members, to foster independence, to renew interest and skills in social relationships, and to decrease dependence on institutional staff.

Once goals have been formulated and the needs, interests, concerns, educational level, and other important characteristics of participants have been determined, specific objectives need to be decided upon. The key learning objectives in a health promotion class take into consideration an assumption of progression of learning. This concept allows for varying learner styles and the ability to integrate and apply new information. By setting objectives along this continuum of learning, participants can progress at their own rate and use their own style.

This progression approximates these steps: increased awareness, understanding, decision making, experimentation, and evaluation. Using stress management as an example, the five objectives for learning include:

1. Increasing awareness: Can I recognize when I am experiencing stress? Am I aware of my bodily responses and my emotional responses? Can I identify situations and events (called stressors) that elicit a stress response in me? Do I notice that individuals experience stressors differently?
2. Understanding: Am I aware of the relationship between life changes and stress? Do I accept stress as a natural part of all life but realize stress overload (distress) can lead to malfunction, disease, exhaustion, and even death? Do I understand that deprivation, or understimulation, as well as overstimulation, frustration, and adaptation to change are factors that cause stress? Am I aware that skills can be learned to induce deep relaxation and that the effectiveness of these skills increases with regular practice?
3. Decision making: Can I change my attitude, if I so wish, to influence the way I interpret a potentially stressful situation? What value is there for me in learning techniques to modify my response to stressors? What will the benefits and disadvantages be if I attempt to modify undesirable situations that affect me and/or my community?
4. Experimentation: Am I practicing the techniques I learned in class at home? What can I do to encourage and reinforce my motivation to practice?
5. Evaluation: Which technique(s) is (are) the most useful to me? Do I need to make any changes in the technique to meet my needs appropriately? Do I take action in managing stressful situations when possible? Do I use a relaxation technique on a regular basis? Do I take "change of pace" breaks during the day and time off (day outings, vacations) on a regular basis? Do I use a "body scan" to locate and release tension daily?

CREATING A PARTICIPATIVE LEARNING CLIMATE

There are a number of facilitator qualities and skills which help to create a positive climate for risking and sharing. Warmth, sincerity, patience, understanding, flexibility, and perseverance are all human qualities which help to create a climate of openness and trust. Skills in interpersonal communication, listening, and empathizing are also important.

In creating a positive learning climate, a facilitator should do the following:

—Determine the educational level and life experience of the group as a whole, and proceed from there.
—Provide a non-threatening atmosphere which encourages participants to express their ideas, opinions, and personal experiences.
—Provide positive and constructive feedback to each participant; ask for feedback from participants about the facilitator's role.
—Ask questions to stimulate participation in discussion.
—Challenge participants without attacking their self-worth, self-confidence, or integrity.
—Ask questions which orient or reorient participants to the task or topic of discussion.
—Encourage participants to share their resources with one another.
—Recognize when a problem exists in group interaction and determine if it should be discussed with the group or dealt with on an individual basis.
—Be aware of participant body language as a means of conveying involvement or withdrawal in a group environment.
—Be flexible and willing to modify educational plans in order to meet the needs of participants.

FACILITATING INTERACTION AMONG PARTICIPANTS

Much, if not most, of the learning that occurs when using the participative learning approach is the result of interaction among participants. This interaction is most frequently verbal interaction, e.g., discussion. However, behavioral interaction is also an important dimension of learning, such as giving support to one another through hugs and giving each other stress reducing back rubs.

The guidelines for creating a positive learning climate outlined earlier are equally applicable to facilitating interaction. There are several other suggestions for the facilitator to keep in mind. These also have the effect of creating a more positive learning environment.

—Begin sessions with "get acquainted" non-threatening activities and gradually move toward more complex, personally involved activities.
—Give clear, concise instructions for activities.
—Offer praise, encouragement, and support for group members' contributions.
—Help build additional support into the learning process by suggesting that students work with partners to encourage and help each other.
—Have a written agenda and objectives for each meeting posted on a large newsprint paper so that everyone will have a common point of reference.
—Allow time for discussion, clarification, and revision of the agenda.
—Be sensitive to the attention span of participants. Plan short breaks and a variety of activities. During breaks, suggest that participants talk with someone they don't know very well.
—The class sessions pack in a great deal of information. Allow the group to set its own pace for learning new information and trying out new skills.
—Be willing to take risks! When participants demonstrate a readiness for leadership, encourage them to experiment.
—Be enthusiastic about your topic and the activities. It can be contagious!

A WORD ABOUT DISCUSSION

A basic component of most learning activities is discussion. Discussion groups may also be used as the sole activity to attain an objective. However, we know from experience that effective, stimulating discussions do not just happen. Below are ten principles which facilitators should be aware of in conducting effective discussions.

1. It is easier to speak in a smaller group than in a larger one.
   —If participants seem to be reluctant to participate in a discussion, divide into smaller groups no larger than four people.
   —Have each participant present his or her ideas; once they have spoken in small groups, participants will be less reluctant to contribute in a larger group.

2. Participants need time to think.
   —Don't be afraid of silence.
   —Wait rather than rushing to rephrase questions or ask new ones.
3. Participants may feel less inhibited when talking with peers.
   —Encourage participants to facilitate the discussions or assign someone to take over.
4. Emphasize that discussions have no right answer and a conclusion or resolution need not always be reached.
   —When participants realize there is no one right answer, they will be more open about expressing opinions.
5. Participants must feel their opinions are valued.
   —Let them know you are interested in what they think. Be an active listener.
   —Point out important contributions or ideas an individual makes, especially if they are overlooked by the group.
   —React positively to all contributions.
6. Discussions should be relevant to the participants' own lives and concerns.
   —Present questions which will elicit individual concerns and experiences regarding the topic.
7. Participants cannot always be prepared for discussions.
   —Again, be flexible.
   —Find out if there are reasons the group is unprepared for the discussion. There might be circumstances of which you are unaware (deaths, community problems, etc.); these circumstances could be appropriate topics for discussion.
   —Sensitivity to group needs is important.
8. Quiet, shy participants usually need to be specifically encouraged to share their opinions.
   —This is often the best way of drawing them into the conversation. Once they begin talking, they will often volunteer on their own.
   —Try to provide other tasks that allow participants to make a variety of contributions such as recording ideas on a blackboard.
9. Aggressive members may tend to monopolize discussions.
   —Encourage others to participate by asking if they agree or disagree with what has been said.
   —If the discussion remains too one-sided, note this and ask the monopolizer(s) to help by remaining silent for awhile.

10. Physical surroundings can influence the dynamics of a discussion.
    —Sit in a circle or U-shaped formation so everyone can see and hear each other.
    —Meet in a comfortable setting.
    —Avoid noisy surroundings.

Audiovisuals

Some other general guidelines, not specifically for facilitating interaction but important just the same, apply to the use of written instructional materials and audiovisual materials.

—When selecting instructional materials, consider that many older adults have vision and/or hearing impairments.
—Use large-size newsprint paper or a chalkboard to emphasize key concepts. Write legibly and in large print.
—Other visual aids should be presented at a moderate distance and be large enough to be seen by all.
—Have sufficient light but avoid glare.
—Preview all materials such as films, videotapes, film strips, and slides checking for content, accuracy, film and audio quality, and length.
—Audio tracts should be distinctly spoken, low pitched, and free from noise interference.

Educational Activities

Selecting the most appropriate activity for attaining any given objective is not an easy task. There are numerous pitfalls to avoid. One common mistake is to select the activity the facilitator knows best and feels most comfortable with rather than the one that would best meet the objective. Facilitators must work to continually expand their knowledge and skills of various activities so that choices can be made from a broad repertoire of potential activities.

The following activities have all been tried and tested at the Wallingford Wellness Project. A brief description will be given for each activity, followed by potential uses, instructions for use, and any special considerations for the facilitator to keep in mind.

Lecture. A prepared presentation of factual information on a particular theme or topic; covers a great deal of information in a short period of time.

Uses:

—As an introduction to a new topic
—To identify or explore a problem
—As a response to questions raised by participants

Instructions and Considerations:

—Keep in mind the learning goals; decide on what material you need to cover and in what depth.
—Use audio-visual aids whenever possible.
—Keep within the allotted time and allow for questions.
—Time requirements: optimum–20 minutes; maximum–25 minutes; minimum–15 minutes

Symposium. A series of individual presentations by several speakers on various aspects of a problem or topic.

Uses:

—To present a complete picture of a problem or subject
—To break down a complex problem by examining its components, differing viewpoints, alternatives, and consequences

Instructions and Considerations:

—Choose speakers based on the learning objectives of the meeting.
—Meet with the speakers prior to the symposium to agree on a breakdown of the subject, who will cover what topics, order of appearance, and time allotted for each.
—Prepare an outline or agenda of the symposium for the class.
—Time requirements: Per speaker: optimum–15 minutes; maximum–20 minutes; minimum–10 minutes

Panel. A spontaneous discussion on a particular topic or issue by a small selected group of individuals; it may be formal or informal.

Uses:

—To explore a problem or issue
—To get different points of view and to look at a problem from different perspectives
—To impart an understanding of a complex problem

Instructions and Considerations:

—Choose a topic which is timely and of concern to your audience.
—Have the group decide on appropriate panel members to invite. Keep in mind the panel should represent different views and experiences.
—A moderator should be chosen who is impartial, can ask pertinent questions, and can summarize well.
—Select and invite panel members far enough in advance to give them time to make arrangements and prepare for the discussion.
—Seat panel members so they can see each other and see and be seen by the audience. Lively speakers should be seated on the ends with quieter panelists between.
—Time requirements: optimum–30 minutes; maximum–45 minutes; minimum–25 minutes

Brainstorming. Participants join together to generate new ideas, solutions, or alternatives to a problem. Stimulates creative thinking and recognizes the wealth of information members bring to the group situation.

Uses:

—To generate suggestions for topics to be discussed, possible solutions to a common problem, or anything that has a number of possibilities
—To help build group cohesiveness and stimulate participation

Instructions and Considerations

—Present the problem.
—Ask someone to record all the ideas on a sheet of newsprint paper or on a chalkboard.
—Give an explanation of the rules:
 —No criticism of ideas until after the brainstorming;
 —Any idea goes—the crazier the better, it can be tamed down later;
 —The sky is the limit—don't worry about the practicality or feasibility of the idea—you can evaluate it later;
 —Adding on to a previous idea or combining ideas is okay—anything to make an idea better!
—State how long the group will have to brainstorm.

—After brainstorming, discuss the ideas, refine, evaluate, and prioritize so the group comes up with a few main suggestions. Depending on the problem, group consensus for one solution may be desirable.
—Ideas can also serve as points for future discussions or subjects for role plays.
—Time requirements: optimum–15 minutes; maximum–20 minutes; minimum–10 minutes

Readings. Printed materials which provide background information, a variety of viewpoints, or in-depth discussion of topics discussed in group.

Uses:

—To reinforce what is presented or discussed in class
—To provide additional information to those who request it
—To offer another opportunity for participants to share information

Instructions and Considerations:

—Ask everyone to share with the group materials they have read that are relevant to topics being discussed.
—Provide articles, books, and magazines which are thought-provoking and relevant. Have copies available in large print.
—Time requirements: at participants' pace

Audio-visual Materials. Films, slides, videotapes, etc., which are informative as well as entertaining.

Uses:

—To open a discussion; when shown at the beginning of a session, audio-visual materials provide participants with similar information and reference points.
—To close a discussion, lecture, or symposium; helps reinforce concepts presented by providing visual illustrations.
—To provide a situation for role play; by stopping the film before an important point, possible conclusions can be played out.

Instructions and Considerations:

—Have all the necessary equipment and know how to operate it.

—Introduce audio-visuals by title. Demonstrate pertinence to learning goals and indicate what points to look for.
—Follow with appropriate activity.
—Time requirements: varies depending on purpose and use

Buzz Session. A brief, informal discussion of a topic or issue with small sub-groups of a larger participant group. Discussants typically have no prior preparation.

Uses:

—To increase awareness of critical issues prior to a formal presentation such as a lecture or film
—To generate greater interest and involvement with a topic and to increase sensitivity to issues and problems

Instructions and Considerations:

—Briefly introduce the topic to be discussed in general terms. State the objectives of the session, i.e., to identify issues, to specify problems, etc.
—Give procedural instructions—how to break into groups, the time allotted for the activity, etc.
—When time is up, reconvene the larger group. Have someone from each group present his or her group's conclusions.
—Discuss briefly, summarizing the points made.
—Time requirements: optimum–10 minutes; maximum–15 minutes; minimum–5 minutes

Demonstration. An activity where one or more persons, proficient in a specific skill or procedure, demonstrates the skill to the larger group. The lesson should include the "whys" of a demonstrated skill as well as the "hows."

Uses:

—Skills training may involve the use of equipment, for example, in food preparation or blood pressure management.

Instructions and Considerations:

—Provide each participant with a set of instructions you will follow.
—Have the participants examine the equipment and explain how it works.

—Demonstrate the procedure slowly explaining and answering questions as you go along.
—Demonstrate several times before expecting participants to try the procedure or skill on their own.
—Have participants try the skill and provide feedback.
—Time requirements: optimum–15 minutes; maximum–20 minutes; minimum–10 minutes

Case Study. A written vignette describing an incident, event, or situation to be analyzed and discussed. Must contain sufficient detail to make it possible for groups to analyze the problems involved.

Uses:

—To help participants think about issues or problems and how they might act or react in similar situations

Instructions and Considerations:

—Provide a copy of the case to everyone. Ask someone to read it out loud or do so yourself.
—Briefly summarize the situation discussed but avoid going beyond the stated facts.
—Ask someone to analyze the case, point out the problems, or suggest alternative methods of dealing with the situation.
—At no time during the discussion should you direct the course of the discussion or interject personal opinions or solutions.
—At the conclusion of the discussion, summarize the key points.
—Time requirements: optimum–40 minutes; maximum–60 minutes; minimum–30 minutes

Interview. Probing questions asked by selected participants of a responder who may be a professional in a subject area or someone who is role playing.

Uses:

—To gain additional information from someone with expertise in a given area
—To practice skills in formulating and asking questions, for example, questions of your doctor

Introductions and Considerations:

—Have participants select the person or persons to be interviewed and the group members who will conduct the interviews.
—Have participants discuss what they want to know and formulate some questions.
—Have someone make the arrangements or do so yourself.
—Time requirements: optimum–30 minutes; maximum–40 minutes; minimum–20 minutes

Role Play. The acting out of a hypothetical situation by two or more persons. Each person assumes a role and plays the part as he or she feels would be appropriate if it were a "real life" situation.

Uses:

—To try out suggestions or solutions to problems by creating a situation in which they might be used
—To increase involvement and identity with a problem
—To provide an opportunity for individuals to step into someone else's "shoes," thus giving them insight into other points of view

Instructions and Considerations:

—Introduce the problem and briefly outline the situation.
—Ask for volunteers or assign roles.
—Allow time for players to ask questions and internalize their roles.
—Allow the scene to unfold without interruptions.
—Upon conclusion of the scene, open for discussion. Let the actors explain their actions before the audience begins making comments.
—Encourage everyone to share their feelings and reactions.
—Time requirements: optimum–45 minutes; maximum–60 minutes; minimum–30 minutes. Time should include allowances for introduction, role assignments, etc.

Problem Solving. Hypothetical or real problems are assigned to small groups of 3 or 4 participants who work together to reach a collective decision. Hypothetical problems may be constructed so that they can be completed within one session or extended over a longer period of time.

Uses:

—To give participants experience working collectively
—To help participants improve and use their problem-solving skills
—To provide participants with an opportunity to examine a problem more closely

Instructions and Considerations:

—Divide the larger group into smaller sub-groups.
—Provide a written description or outline of the problem to be investigated or have each sub-group choose a topic. Suggestions of possible problem areas may be made.
—Upon completion of the problem-solving sessions, have someone from each group summarize that group's findings and conclusions.
—Discuss the findings as well as problems encountered by the groups in the actual problem-solving process.
—Time requirements: optimum–60 minutes; maximum–90 minutes; minimum–45 minutes. Time varies depending on length and complexity of the problem.

Field Trip. Guided tour of a facility, project, or recreation spot.

Uses:

—To help participants understand the internal structure of an organization
—To observe and compare alternative organizational methods
—To observe people and processes in their work or leisure environments
—To promote group cohesiveness and social interaction

Instructions and Considerations:

—Prior to the outing, discuss why a particular site is being visited and what participants should observe.
—Make the necessary arrangements keeping in mind any disabilities or impairments of your participants.
—After the field trip, discuss participants' observations, if they met their expectations, and how this new information can be used.
—Time requirements: optimum–1½ hours; maximum–3 hours; minimum–30 minutes (not including travel time)

Educational Tools for Health Promotion Activities

Three tools are intended for use in all four health promotion content areas. The first, the "Affirmation of Health" forms help participants begin and maintain the process of making changes that contribute to a healthier lifestyle. The "Evaluation" and "Assessment of Participation" forms are tools for improving the program so that it will become more effective in facilitating positive lifestyle changes for future participants.

"Affirmation of Health." The "Affirmation of Health" is an agreement made with oneself for health-enhancing changes in one's lifestyle. A separate form is used in each of the four program areas. (In the nutrition component it may be substituted with a self-contract form based on the same theories, but tailored specifically to nutrition-related goals.) It is designed specifically to help the person:

—identify a long-term personal goal in the program area;
—identify specific activities (short-term goals) necessary to reach the goal; and
—design a multi-faceted plan for reinforcing his/her efforts.

The effectiveness of this tool is a function, in part, of the nature of the goals. All goals should be realistic, easily attainable, clear, stated in positive terms, and reflect a pace suitable to the individual.

The plan for reinforcing the person's efforts (item 3 on the form) is a second determinant of this tool's effectiveness. The form contains five reinforcers that the individual selects to help him/her make positive changes. The "self-reward" is highly recommended for its power of nurturing a sense of achievement.

A third factor influencing the effectiveness of the "Affirmation of Health" form is in-class peer support. One way to achieve this support is the "sharing" time at the beginning of each class session. During the sharing, participants are encouraged to relate to the group their experiences in carrying out the week's activity (short-term goal).

The facilitator should recognize *all* efforts shared, regardless of success or failure. Change takes time and the very process of *trying* is valuable. The facilitator should encourage participants to recognize and support each other's efforts during sharing (e.g., clapping for a success or offering to work with a classmate having difficulty). A second mechanism for building peer support is to fill out the forms in pairs or small groups, with participants agreeing to help each other during the week (e.g., brisk walking together or telephoning to offer encouragement).

Finally, this tool can be effective only if it is clearly and thoroughly explained by the facilitator at the time of its introduction.

Evaluation. In the class sessions, participants are given the Evaluation form during the final session of each program area and are asked to complete it at that time. They should be encouraged to be honest and thorough in their responses. The facilitator should explain that participants' feedback is a highly important means of continually improving the course for future participants.

The intent of this form is twofold. First, participants' comments are used to modify materials, activities, course content, and facilitators' styles. Second, enabling participants to see that they have a role in shaping this program allows them to recognize another small yet significant indication of their own power to influence their environment. The Evaluation form contributes to the overall program goal of empowering people to assume more control over their health and wellbeing, and the health of their environment.

Assessment of Participation. The Assessment of Participation form can be given to participants together with the Evaluation at the final session of the fourth program area. Its use and intent are the same as that of the Evaluation with one exception: participants should be directed to respond to it with regard to your entire health promotion program, rather than just one of the four core content areas.

AFFIRMATION OF HEALTH

Name______________________________

1. The major goal I want to accomplish in the next ______ weeks is:

2. How I plan to accomplish this goal is:

Week	Activity	Where	When	Times Per Week	With Whom

3. In order to help me follow through with my activity, I will:

   ______Keep a journal of my thoughts, feelings, and reactions.

   ______Keep records of "daily activities log."

   ______Invite a friend or family member to work with me in my effort to change: The way this person will help me is:

   ____

   ____

   ______Invite someone from the class to work with me in my effort to change: The way this person will help me is

   ____

   ____

   Reward myself by______________________________

   ____

EVALUATION

Would you please help us evaluate our classes and teaching by filling out this questionnaire? We appreciate all of the information you can give us. Thank you.

____Female ____Male Age:_____

1. I learned from my participation in the ________________ class:
   A. ______A whole new approach to life
   B. ______Some new approaches to my lifestyle
   C. ______How to better integrate what I already know
   D. ______Not very much
   E. ______Other (please be specific)______________________________

   ____

2. The materials used in the class: (check as many as are appropriate)
   A. ______Needed more explanation
   B. ______Were new to me
   C. ______Helped me to achieve my goals in lifestyle change
   D. ______Were not very useful
   E. ______Other (please be specific)______________________________

   ____

   ____

3. The activities in class: (check as many as are appropriate)
   A. ______Needed more explanation
   B. ______Were new to me
   C. ______Helped me to achieve my goals in lifestyle change
   D. ______Helped me increase my knowledge
   E. ______Were not very useful
   F. ______Other (please be specific)______________________________
   ____
   ____
   ____

4. I really like it when the class leaders:
   ____
   ____
   ____

5. Some things that the class leaders could do to improve their presentations are:
   ____
   ____
   ____

6. Two things that I enjoyed most about the class are:
   ____
   ____
   ____

7. Two things that I would do to improve the class are:
   ____
   ____
   ____

8. Because of the class, specific lifestyle changes I have made include:
   ____
   ____
   ____

9. The most important information I have learned from this class is:
   ____
   ____
   ____

Name ____________________
Class ____________________
Date ____________________

ASSESSMENT OF PARTICIPATION

I. Participation in the program can occur in many ways. After each item below, please circle the number that best reflects your level of participation. Number 1 represents "no participation" and number 5 represents "highest participation."

A. Listening in Classes.

1 2 3 4 5

B. Discussing and sharing my ideas in class.

1 2 3 4 5

C. Asking questions in classes.

1 2 3 4 5

D. Doing in-class activities, such as going on field trips, or doing stretching exercises.

1 2 3 4 5

E. Doing out-of-class activities, such as reading about nutrition, asserting myself, doing aerobic exercises like brisk walking, or making a meatless meal.

1 2 3 4 5

F. Getting to know other class members during class time.

1 2 3 4 5

G. Getting together with other class members outside of class time by calling or visiting.

1 2 3 4 5

H. Sharing with others (family, friends, acquaintances) what I have learned in these classes.

1 2 3 4 5

II. Please circle the number below that best reflects your *overall* participation in the program. Number 1 represents "no participation" and number 10 represents "highest participation."

1	2	3	4	5
6	7	8	9	10

Resources

It is not within the scope of this manual to provide the reader with a thorough discussion of adult learning and group dynamics. We encourage you to explore other resources. To get you started we offer the following:

Beal, G. M. et al. *Leadership and dynamic group action.* Ames, IA: The Iowa State University Press, 1962.

This is an excellent book which covers the dynamics of group interaction, methods and techniques, and ways of evaluating the group process and its progress. Written with a sense of humor and conscious avoidance of jargon, this resource is understandable and easy to read.

Burnside, I. *Working with the elderly: Group process and techniques.* North Scituate, MA: Duxbury Press, 1978.

This book covers the basics of working with groups, but focuses primarily on therapeutic groups for institutionalized older adults. Offering a wide variety of strategies and techniques, this source is a must for anyone working with older adults.

Jacobs, B. *Senior centers and the at-risk older person.* National Institute of Senior Centers, Washington, D.C.: The National Council on the Aging, Inc., 1980.

Kemp, G. E. *Instructional design: A plan for unit and course development* (2nd Ed.). Belmont, CA: Fearon-Pitman Publishers, Inc., 1977.

Written for teachers in primary, secondary, or college settings, the author presents an easy-to-follow method for planning educational objectives.

Mager, R. F. *Preparing instructional objectives* (2nd Ed.). Belmont, CA: Fearon-Pitman Publishing, Inc., 1975.

This resource is a "how to" book for developing and stating clear, concise learning objectives. Presented in a unique format, the author actively involves the reader in recognizing the essential characteristics of a useful objective.

Miller, I., & Solomon, R. The development of group services for the elderly. *Journal of Gerontology Social Work*, 1980, *2*(5), 241-257.

This article discusses reasons for developing group services for older adults, types of groups, and offers valuable guidelines for facilitators.

Nicoley-Colquitt, S. Preventive group interventions for elderly clients: Are they effective? *Family and Community Health*, 1981, *3*(4), 67-68; 80-85.

This article reviews various group strategies used with older adult populations with a focus on prevention. It also examines the benefits of these strategies and makes recommendations for further research in this area.

Whitbourne, S. D., & Sperbeck, D. J. Health care maintenance for the elderly. *Family and Community Health*, 1981, *3*(4), 11-27.

This book provides an excellent discussion of the learning process and what to be aware of when helping an older person learn new coping mechanisms.

Stress Management

In this section you will find information, suggested activities, and resources for a basic stress management course.

Who can help you develop and deliver the class? We recommend stress management specialists, therapists, and counselors from:

—Community mental health clinics
—Local Mental Health Association
—YMCA or YWCA
—City or County Parks and Recreation
—Licensed massage therapists
—Local hospital health education departments

Helpful Publications

Help Yourself (free) is available from:

Blue Cross Association
840 North Lake Shore Drive
Chicago, IL 60611

Other information and educational materials in the area of stress management and mental health are available from:

National Association for Mental Health
1800 North Kent Street
Arlington, VA 22209

National Clearinghouse for Mental Health Information
Public Inquiry Section
National Institute of Mental Health
5600 Fishers Lane
Rockville, MD 20857

Additional readings are suggested at the end of this section.

SESSION 1: AN OVERVIEW OF STRESS AND STRESS MANAGEMENT

Session Agenda

1. Welcome participants, preview the agenda for the session, and make announcements.
2. Facilitate discussion of participants' expectations of the stress management course, explain the goals of the course, and familiarize participants with what the course will include.
3. Define stress.
4. Describe the body's reactions to stress and relaxation.
5. Identify four main effects of stress.
6. Teach a relaxation technique: deep breathing.
*7. Identify numerous ways of coping with stress, and facilitate a group discussion regarding which of these to address in the course.
8. Evaluate the session.
9. Preview the agenda for Session 2.

NOTE: This plan assumes that participants and facilitator have met. If this is not the case, these activities should be preceded by introductions. Please refer to The Personal and Community Self-Help component, Session 1, Activity 1, for an introductory game.

Handouts

Handout 1A - Goal of the Class
Handout 1B - Bodily Reactions of Stress and Relaxation

Materials Needed

Flip chart
Magic markers
Session agenda printed on flip chart
Session 2 agenda

---

*This module provides course content and format for four class sessions. In order to further the overall aim of increasing participants' sense of control over their own wellbeing, it is suggested that they determine the specific stress management techniques to be addressed in subsequent sessions—if the course indeed contains more than four sessions. If the fourth is the final session, then the second portion of Activity 7 should be deleted.

Activity 1: WELCOME PARTICIPANTS, PREVIEW THE AGENDA FOR THE SESSION AND MAKE ANNOUNCEMENTS.

Methodology: Flip chart

After welcoming participants, direct their attention to the agenda printed on the flip chart. Briefly list each item. Make any necessary announcements, and invite participants to do likewise.

Activity 2: FACILITATE DISCUSSION OF PARTICIPANTS' EXPECTATIONS OF THE STRESS MANAGEMENT COURSE, EXPLAIN THE GOAL OF THE COURSE, AND FAMILIARIZE PARTICIPANTS WITH WHAT THE COURSE WILL INCLUDE.

Methodology: Group discussion
Lecturette
Handout 1A

Ask the participants what they hope to gain from the course. If necessary, stimulate discussion by rephrasing the question; ask why they are taking the course, or what changes they would like to make in their lives regarding stress management.

Refer participants to Handout 1A. Read it aloud or paraphrase its contents.

Activity 3: DEFINE STRESS

Methodology: Group discussion

Ask the group for definitions of stress. Then give Seyle's definitions as follows:

"Stress is the nonspecific response of the body to any demand made upon it."
"Stress is not merely nervous tension."
"Stress is not always the nonspecific result of damage."
"Stress is not something to be avoided."
"Complete freedom from stress is death."
A stressor is "the life event or stimulus." (Selye, 1974)

Activity 4: DESCRIBE THE BODY'S REACTIONS TO STRESS AND RELAXATION.

Methodology: Lecturette
Group Discussion
Handout 1B

Paraphrase the contents of Handout 1B. Refer them to the handout for further use. Encourage discussion of their experience with stress reactions and relaxation reactions.

Activity 5: IDENTIFY FOUR MAIN EFFECTS OF STRESS.

Methodology: Group brainstorming

Ask the group to brainstorm some of the effects of stress. Write these on the flip chart, then categorize all items into the four main areas listed below.

1. *Physical:* hypertension (high blood pressure), stroke, arthritis, respiratory disorders, infection.
2. *Emotional:* anxiety, fear, anger, resentment, depression, guilt feelings or helplessness and inadequacy.
3. *Mental:* neuroses and psychoses, manic-depressive illness, schizophrenia, obsessive-compulsive behavior, and hysterical reactions.
4. *Self-destructive habits:* overeating, smoking, overuse of alcohol and drugs.

Activity 6: TEACH A RELAXATION TECHNIQUE: DEEP BREATHING.

Methodology: Lecturette
Demonstration and practice
Group discussion

The goal of the course is to learn techniques of managing unwanted stress to that undesirable effects of stress may be minimized.

''Deep breathing'' or ''abdominal breathing'' can be used in nearly any setting. Deep breathing relaxes the muscles and allows more oxygen to flow through the body. It should be practiced 5-10 minutes each day. It is particularly useful when one is feeling tense.

Demonstrate deep breathing as follows:

—Sit in a comfortable position, with feet flat on the floor and arms resting on your lap.
—Inhale slowly through the nose, first filling the abdomen with air, then the chest.
—Hold your breath for the count of 5.
—Slowly exhale through the nose, pulling in your abdomen, thus allowing your lungs to empty of air.

Repeat the process. Ask participants to join you.

Briefly brainstorm with the group tense situations in which deep breathing could be a practical and effective tool. For example, sitting in the dentist's chair, or waiting to give a speech.

Activity 7: IDENTIFY WAYS OF COPING WITH STRESS AND FACILITATE A GROUP DISCUSSION REGARDING WHICH OF THESE TO ADDRESS IN THE COURSE.

Methodology: Group discussion
Flip chart

Divide into small groups of approximately four persons. Discuss ways each participant has handled stress in the past. Ask each small group to appoint a recorder/spokesperson who will report to the large group. Reconvene and ask the spokespersons to report on the techniques mentioned in their small groups. Write all items on the flip chart. Allow time for discussion and questions.

If the stress management course will last more than four sessions, ask the group to determine which techniques they would like to address in the sessions remaining after the fourth session. Encourage them to decide upon a decision-making process. If they do not do so, suggest eliminating any items on the list which are not feasible, and voting on those that remain. The facilitator should participate by eliminating any items for which he/she would not be able to arrange resources.

Activity 8: EVALUATE THE SESSION.

Methodology: Group discussion
Flip chart

The purpose of the evaluation is to continually improve the courses by incorporating participants' feedback and perceptions. To elicit evaluatory comments, ask the group, "What were the most useful aspects of this session?" and "How could the session have been improved?" It is important that the facilitator be receptive to constructive criticism.

Activity 9: PREVIEW THE AGENDA FOR SESSION 2.

Methodology: Flip chart

The agenda for Session 2 should already be written on the flip chart. Briefly read it aloud.

HANDOUT 1A

GOAL OF THE CLASS

The overall goal of this class is to help yourself manage unwanted stress in your life.

You will learn to identify stress, to recognize stressful situations (stressors), and to manage stress by practicing and applying relaxation techniques.

Relaxation techniques that may be taught in this class include:

Counting Breaths[1]
Deep Breathing
Shoulder and Neck Massage
Progressive Relaxation
Instant Relaxation Exercise[2]

With the help of "Affirmations of Health" and a daily "Stress and Tension Log," you will be able to use these techniques in your day-to-day life. As a result, you may feel more relaxed and better able to manage stress.

We value your active participation in the group and hope you will help us create an enjoyable and lively class.

---

[1]Adapted from McCamy, J., & Presley, J. *Human Lifestyling.* New York: Harper & Row, 1975.

[2]Adapted from Farquhar, J. *The American Way of Life Need Not Be Hazardous to Your Health.* New York: W. W. Norton, 1978.

HANDOUT 1B

BODILY REACTIONS OF STRESS AND RELAXATION

Everyone experiences stress in a variety of different situations. There are, however, a common set of bodily responses to "stressors" or stressful events. For example:

—Nervous system activity increases
—Blood pressure goes up
—Muscles become more tense
—Need for oxygen increases and breathing rate goes up

A "relaxation response" has the opposite effect. By engaging the relaxation response, the nervous system slows down. Blood pressure decreases as does muscle tension and rate of breathing.

You can learn to call forth a relaxation response through stress management exercises like deep breathing and progressive relaxation.

With practice of stress management techniques, you can develop skills to gain greater control of your body and mind and increased ability to regulate the undesirable responses you experience. You can enjoy increasing wellness by incorporating stress management techniques into your developing "healthstyle."

SESSION 2: BASIC STRATEGIES TO IDENTIFY AND MANAGE STRESS IN YOUR LIFE

Session Agenda

1. Preview the agenda for the session and make announcements.
2. Review content of Session 1.
3. Explain the steps in learning to manage stress.
4. Enable participants to identify personal stress levels.
5. Teach a stress management technique: shoulder and neck massage.
6. Guide participants in developing individualized stress management action plans.
7. Introduce the "Daily Stress and Tension Log."
8. Evaluate the session.
9. Preview the agenda for Session 3.

Handouts

Handout 2A – Simplified Self-Scoring Test for Gauging Stress and Tension Levels
Handout 2B – Observer Behavior Rating Inventory
Handout 2C – Shoulder and Neck Massage
Handout 2D – Affirmation of Health (sample)
Handout 2E – Affirmation of Health
Handout 2F – Daily Stress and Tension Log

Materials Needed

Flip chart
Magic markers
Session agenda printed on flip chart
Agenda for Session 3 printed on flip chart

Activity 1: PREVIEW THE AGENDA FOR THE SESSION AND MAKE ANNOUNCEMENTS.

Methodology: Flip chart

Read through the agenda items printed on the flip chart. After making announcements, invite participants to make announcements they may have.

Activity 2: REVIEW CONTENT OF SESSION 1.

Methodology: Small groups

Explain that this review will take place in small groups. Request the group to break into groups of three. Each person should do the following aloud:

1. Choose a stressful situation which occurred during the past week.
2. Describe the situation and his/her response—bodily and/or emotional—to the situation.
3. Describe what he/she did to cope with the situation and, if this was a stressful response, what could have been done to elicit a relaxation response in this situation.

Facilitator should role-play the procedure before participants begin.

(Reconvene the large group before proceeding to the next objective.)

Activity 3: EXPLAIN THE STEPS IN LEARNING TO MANAGE STRESS.

Methodology: Lecturette

Explain to the group the following:

Managing stress is one way you can take control of your life. To bring about positive lifestyle changes regarding stress, you may go through a series of learning steps. Each of you will progress at your own rate. Many of the exercises described in this course can be used by special populations, e.g., the institutionalized elderly and those who are homebound.

There are six steps which have been identified in this learning progression:

1. Identifying stress problems.
2. Increasing awareness of stress sources and stress responses.
3. Building confidence and commitment to change.
4. Developing a stress management action plan.
5. Evaluating the stress management plan.
6. Maintaining positive changes.

These steps do not occur one at a time; they overlap and recur continually. Each of them will be addressed in this Stress Management course. We began to learn some of them in Session 1.

Activity 4: ENABLE PARTICIPANTS TO IDENTIFY PERSONAL STRESS LEVELS.

Methodology: Lecturette
Handout 2A
Handout 2B

A significant aspect of identifying one's stress problems is identifying general stress levels. We will now seek to do so by completing the "Simplified Self-Scoring Test for Gauging Stress and Tension Levels," and by asking a friend to complete the "Observer Behavior Rating Inventory."

Distribute the first of these two questionnaires (Handout 2A), and ask each participant to complete the form in class. After these have been completed, pass out the second questionnaire (Handout 2B). Request that each person ask a friend to complete it. The results of this second, more objective inventory can be used to check those of the first. All results will be kept confidential; they are for use by the participant only.

Activity 5: TEACH A STRESS MANAGEMENT TECHNIQUE: SHOULDER AND NECK MASSAGE.

Methodology: Demonstration and practice
Handout 2C

Request each person to pair up with someone. Choose one person so you can demonstrate the technique. Lead the group through the technique as outlined in Handout 2C. Participants should not merely observe, but also *do* the massage. Ask partners to switch places and repeat the process. Facilitator may choose a different person to work on.

It is important to teach this technique at a relaxed pace and in a relaxed tone of voice. Be aware that some persons may prefer not to touch or be touched; respect this preference.

Activity 6: GUIDE PARTICIPANTS IN DEVELOPING INDIVIDUALIZED STRESS MANAGEMENT ACTION PLANS.

Methodology: Lecturette
Handout 2D
Handout 2E
Small groups

The next steps in learning to manage stress are to build confidence and commitment to change and to develop a stress management action plan. The tools for carrying out these steps will be the "Affirmation of Health" form (Handout 2E) and the "Daily Stress and Tension Log" (Handout 2F). Refer the group first to the "Affirmation of Health" form and the sample "Affirmation of Health" (Handout 2D); elaborate upon them as follows.

Use of the "Affirmation of Health" is based upon setting appropriate goals and using peer support systems. Each person will establish a major (or long-range) goal similar to that recorded on Handout 2D, item #1: "To learn how to relax and better manage the stress in my life.'' The key to realizing this goal is the weekly activity (or short-range goal) decided upon and recorded in item #2 of the form. It should:

—be easily attainable;
—use techniques learned in class;
—be relevant to the individual

Peer support and an appealing self-reward are keys to obtaining short-term goals.

Instruct participants to break into groups of three, and develop their individualized action plans. They can help each other decide on short-range goals for the week, as well as plans for peer support and self-rewards. Suggest that this week's goal include either of the two techniques learned in Session 1 and 2 (deep breathing or shoulder and neck massage).

Activity 7: INTRODUCE THE ''DAILY STRESS AND TENSION LOG.''

Methodology: Handout 2F
Lecturette
Group discussion

Refer participants to Handout 2F, the Daily Stress and Tension Log. It is a supplemental tool for developing action plans for stress management. By filling it out each week, one may increase awareness of individual stress patterns, and use this awareness in identifying which stressors to address in the following week's short-term goal.

To increase participants' level of comfort with this log, provide an example for one day; then ask one or two participants to provide examples.

Encourage any questions regarding use of the log. Request participants to make use of it during the following week and to bring it to Session 3 where it will be used in filling out the ''Affirmation of Health'' for the week.

Activity 8: EVALUATE THE SESSION.

Methodology: Group discussion

Same as Session 1, Activity 8.

Activity 9: PREVIEW THE AGENDA FOR SESSION 3.

Methodology: Flip chart

Same as Session 1, Activity 9.

HANDOUT 2A

SIMPLIFIED SELF-SCORING TEST FOR GAUGING STRESS AND TENSION LEVELS
(Circle the appropriate number for each item)

BEHAVIOR	OFTEN	A FEW TIMES A WEEK	RARELY
1. I feel tense, anxious, or have nervous indigestion.	2	1	0
2. People at work/home make me feel tense.	2	1	0
3. I eat/drink/smoke in response to tension.	2	1	0
4. I have tension or migraine headaches, or pain in the neck or shoulders, or insomnia.	2	1	0
5. I can't turn off my thoughts at night or on week-ends long enough to feel relaxed and refreshed the next day.	2	1	0
6. I find it difficult to concentrate on what I'm doing because of worrying about other things.	2	1	0
7. I take tranquilizers (or other drugs) to relax.	2	1	0
8. I have difficulty finding enough time to relax.	2	1	0
9. Once I find the time, it is hard for me to relax.	2	1	0
10. My day is made up of many deadlines.	2	1	0

MAXIMUM TOTAL SCORE = 20

MY TOTAL SCORE ________

ZONE	SCORE	TENSION LEVEL
A	14-18	Considerably above average
B	10-13	Above average
C	6-9	Average
D	3-5	Below average
E	0-2	Considerably below average

HANDOUT 2A (CONTINUED)

All individuals but those in Zone E have something to learn about stress management. Nevertheless, all individuals, including those in Zone E, should have another opinion. Ask a friend or relative who knows you well to rate your stress level from his or her observations. Have him or her take the test on Handout 2B, adapted from one developed by Virginia Price and used in the Stanford Heart Disease Prevention Program. This "observer rating" of how you handle stress may give you some surprising and useful insights.

HANDOUT 2B

OBSERVER BEHAVIOR-RATING INVENTORY

Circle the number in the box that most accurately describes ____________.
Use scoring system from Handout 2A.

BEHAVIOR	NEVER	SELDOM (1-2 times a week)	OFTEN (almost every day)	VERY FREQUENTLY (at least once a day)
1. Hurriedness: eats and/or moves fast.	0	1	2	3
2. Talking: speaks fast, in an explosive manner, repeats self unnecessarily, and/or interrupts others	0	1	2	3
3. Listening: has to have things repeated apparently because of inattentiveness.	0	1	2	3
4. Worries: Expresses worries about trivia and/or things he/she can do nothing about.	0	1	2	3
5. Anger/Hostility: gets mad at self and/or others.	0	1	2	3
6. Impatience: tries to hurry others and/or becomes frustrated with own pace.	0	1	2	3

MAXIMUM TOTAL SCORE = 18 SUBJECT'S TOTAL SCORE ________

Reprinted from Farquhar, J. The American Way of Life Need Not Be Hazardous to Your Life. New York: W.W. Norton, 1978.

HANDOUT 2C

SHOULDER AND NECK MASSAGE

Have your partner sit comfortably in a chair. Stand behind him/her and suggest that he/she remove glasses, close their eyes, and think, "I am relaxing."

1. Allow your hands to greet your partner by placing them warmly on his/her shoulders.
2. Apply gentle but firm and even pressure with your thumbs across the top of the shoulders. Work your way toward the neck and then back across to the ends of the shoulders.
3. Using both hands, massage across the top of the shoulders with a kneading motion.
4. Locate the vertebrae at the base of the neck. Place your thumbs on either side of the vertebrae and apply gentle but firm pressure away from the spine. Continue down the back. Do *not* press on the spine itself.
5. Locate indentations at the base of the skull on either side of the spine at the back of the head. Apply rotating pressure with your thumbs.
6. Stand beside your partner. Place one hand on his/her forehead and one hand behind his/her head for support. *Slowly and carefully* rotate the head first in one direction and then the other.
7. Stand behind your partner again and use three fingers to massage the jaw area. Have your partner clench his/her jaw. You will easily find the muscles that need to be rubbed!
8. Use fingers to gently massage the temples. Work across the forehead and back to the temples.
9. Bring your hands back to the shoulders and allow your hands to say "goodbye."

How do you feel? Sharing this exercise with someone is a warm and caring exchange. It is relaxing for both the giver and the receiver of the massage. Some of the steps (3, 5, 7 and 8) can also be used for self-massage.

HANDOUT 2D

AFFIRMATION OF HEALTH

NAME Randall Stevens

1. The major goal I want to accomplish in the next 4 weeks is:
To learn how to relax and better manage the stress in my life

2. How I plan to accomplish this goal is:

Week	Activity	Where	When	Times Per Week	With Whom
1	Massage	Home	Evening	Daily	Wife
2	Deep Breathing	At Work	Afternoon	Daily	Self
3	Counting Breaths	Home	Evening	Daily	Self
4	Counting Breaths	Home	AM & PM	Twice Daily	Self

3. In order to help me follow through with my activity, I will:

_____ Keep a journal of my thoughts, feelings, and reactions.

✓ Keep records of "Daily Activities Log."

✓ Invite a friend or family member to work with me in my effort to change: the way this person will help me is:
To encourage daily practice & to comment upon or ask about my progress

✓ Invite someone from the class to work with me in my effort to change: the way this person will help me is:
To call me every Monday evening to offer encouragement

✓ Reward myself by Taking a two-day fishing trip at Lake Serene

HANDOUT 2E

AFFIRMATION OF HEALTH

NAME________________________

1. The major goal I want to accomplish in the next ______ weeks is:

2. How I plan to accomplish this goal is:

Week	Activity	Where	When	Times Per Week	With Whom

3. In order to help me follow through with my activity, I will:

______Keep a journal of my thoughts, feelings, and reactions.

______Keep records of ''Daily Activities Log.''

______Invite a friend or family member to work with me in my effort to change: The way this person will help me is:
____
____

______Invite someone from the class to work with me in my effort to change: The way this person will help me is:
____
____

______Reward myself by________________________________
____

HANDOUT 2F

DAILY STRESS AND TENSION LOG*

DAY OF WEEK	DATE AND TIME OF DAY	STRESSFUL EVENT (STRESSOR)	MY PHYSICAL RESPONSE	THOUGHTS AND FEELINGS I EXPERIENCED	WHAT I DID
MONDAY					
TUESDAY					
WEDNESDAY					
THURSDAY					
FRIDAY					
SATURDAY					
SUNDAY					

*Reprinted from Farquhar, J. The American Way of Life Need Not be Hazardous to Your Health. New York: W.W. Norton, 1978.

SESSION 3: RELAXATION TECHNIQUES

Session Agenda

1. Preview the agenda for the session, and make announcements.
2. Briefly review content of Sessions 1 and 2.
3. Facilitate sharing of the Stress Management Action Plan recorded in the "Affirmation of Health" forms.
4. Teach a relaxation technique:
   A. Progressive relaxation *or*
   B. Counting breaths *or*
   C. Instant relaxation drill
5. Briefly discuss other stress management techniques.
6. Guide participants in developing their individualized stress management action plans.
7. Evaluate the session.
8. Review the agenda for Session 4.

Handouts

Handout 3A - Relaxation Technique—Progressive Relaxation
Handout 3B - Relaxation Technique—Counting Breaths
Handout 3C - Relaxation Technique—Instant Relaxation Drill

Materials Needed

Flip chart
Magic markers
Exercise materials
Session agenda printed on flip chart
Agenda for Session 4 printed on flip chart
Flip chart page from Session 1, Activity 7

Activity 1: PREVIEW THE AGENDA FOR THE SESSION AND MAKE ANNOUNCEMENTS.

Methodology: Flip chart

Same as Session 2, Activity 1.

Activity 2: BRIEFLY REVIEW THE CONTENT OF SESSIONS 1 AND 2.

Methodology: Group discussion
Lecturette

Ask the group to list the main points that were covered in Sessions 1 and 2. List these on the flip chart as they are mentioned. Use this input to briefly reconstruct the first two sessions. Add to the list any major points not listed by the group. Be sure to include:

—the bodily, emotional, mental, and behavioral effects of stress;
—some ways to cope with unwanted stress;
—recognizing stress levels and life events that cause stress;
—the two stress management techniques that were introduced: deep breathing (abdominal breathing) and shoulder and neck massage; and
—the importance of incorporating stress management into lifestyle.

The intent of this review is not to go into an in-depth discussion of any of these points; rather it is to provide a framework for the content of Session 3.

Activity 3: FACILITATE SHARING OF THE STRESS MANAGEMENT ACTION PLANS RECORDED ON THE AFFIRMATION OF HEALTH FORMS.

Methodology: Handout 2F
Sharing moderated by facilitator

Refer participants to the Affirmation of Health forms (Handout 2E) completed in Session 2. Encourage individuals to share their self-contract for the week. Recognize all efforts made, including unsuccessful attempts.

Herein lies an opportunity for further peer support: When a participant discloses having had difficulty in realizing his/her plan, attempt to draw helpful suggestions or similar experiences from others.

The effectiveness of the Affirmation of Health in enabling long-term lifestyle change depends upon:

1. participants feeling reinforced for developing a stress management action plan and carrying it out; and

2. support and encouragement developing among individuals and their families, friends, and classmates. The facilitator should moderate the sharing so participants interact more with each other than with the facilitator.

Activity 4A: TEACH A RELAXATION TECHNIQUE: PROGRESSIVE RELAXATION.

Methodology: Lecturette
Practice
Handout 3A

A number of stress management techniques enable one to relax. Deep breathing and massage were practiced in the last session. Here, three more relaxation techniques are presented. Choose one to practice in this session.

The first is progressive relaxation. Background information on this technique can be found in *You Must Relax* by E. Jacobson.

Lead the group in practicing the technique. Follow the instructions provided in Handout 3A. During all relaxation exercises a comfortable, quiet environment is essential. The facilitator should maintain a calm, relaxing tone of voice while leading the group through the technique.

Refer participants to Handout 3A for their use in practicing the technique at home.

OR

Activity 4B: TEACH A RELAXATION TECHNIQUE: COUNTING BREATHS.

Methodology: Practice
Handout 3B

Lead the group in practicing the technique, following the instructions provided in Handout 3B.

Refer participants to Handout 3B for practicing the technique at home.

OR

Activity 4C: TEACH A RELAXATION TECHNIQUE: INSTANT RELAXATION DRILL.

Methodology: Practice
Handout 3C

Lead the group in practicing the technique, following the instructions provided in Handout 3C.

Refer participants to Handout 3C for use in practicing the technique at home.

Activity 5: BRIEFLY DISCUSS OTHER STRESS MANAGEMENT TECHNIQUES.

Methodology: Group discussion
Flip chart

The techniques learned so far are only some of many effective relaxation exercises. Relaxation exercises are indeed only one of many ways to manage stress. Ask the group to list other relaxation techniques. Write these on the flip chart.

Place relaxation techniques in the broader context of stress management by returning the flip chart to the list generated in Session 1, Activity 7.

Activity 6: GUIDE PARTICIPANTS IN DEVELOPING THEIR INDIVIDUALIZED STRESS MANAGEMENT ACTION PLANS.

Methodology: Small groups
Handout 2E
Handout 2F

Refer participants to their Affirmation Health forms. Break into groups of three and identify and record a stress management goal for the coming week. Suggest that they:

1. refer to their Daily Stress and Tension Logs to help decide what specific stressor to address;
2. plan to use the techniques learned in today's session; and
3. decide upon realistic, attainable goals.

Activity 7: EVALUATE THIS SESSION.

Methodology: Group discussion

Same as the evaluation activity in Sessions 1 and 2.

Activity 8: PREVIEW THE AGENDA FOR SESSION 4.

Methodology: Group discussion

Same as the final activity in Session 1 and 4.

HANDOUT 3A

RELAXATION TECHNIQUE—PROGRESSIVE RELAXATION

Lie comfortably on your back in a quiet place. Allow yourself to become passive.

Begin by taking a few deep breaths and then relaxing into your natural breathing rhythm.

Tense and release groups of muscles one at a time. Begin with your feet, tense the muscles, hold for a count of five and then release. Move up to your lower legs. Tense, hold, and release. Continue to move up your body—upper legs, buttocks, abdomen, chest, shoulders, arms, hands, and face.

Notice what the tension feels like as you contract each muscle group. Focus on the experience of *letting go* of this tension as you progressively relax parts of your body. Allow the tension to float out of your muscles as you let them go as limp as you can.

Practice this relaxation technique once or twice a day for 5-10 minutes. It is a wonderful inducement to sleep as well as a means of releasing excess muscle tension.

HANDOUT 3B

RELAXATION TECHNIQUE—COUNTING BREATHS

Make yourself comfortable in your chair. Keep your back straight. Try not to move during the exercise. Place your hands on your thighs or in your lap with your thumbs together, providing some tension at that spot so you won't go to sleep. Close your eyes.

Take a deep, slow breath. As you inhale, count '*one*,' to yourself. Then, slowly, exhale all the way out, counting '*two*' to yourself silently. Another inhale is '*three*,' out is '*four*.' Quiet, slow breaths (your breathing will gradually become more shallow). In on '*five*,' out on '*six*.' In, '*seven*,' out, '*eight*.' In on '*nine*,' out on '*ten*.' Keep counting. When you get to '*ten*,' start over at '*one*.' Slow, natural breaths now.

If you lose count, start over at '*one*.' Just count your breaths. If a thought comes, don't let it take hold. Let it pass on through as if it were a gentle breeze passing through your hair. Just count your breaths.

Begin practicing this technique 5 to 10 minutes a day and gradually increase the time to suit your needs. When you finish this exercise, ask yourself, 'how do I feel?'

Adapted from: McCamy, J. *Human Life Styling*. New York: Harper & Row, 1975.

HANDOUT 3C

RELAXATION TECHNIQUE—INSTANT RELAXATION DRILL

Position yourself comfortably either sitting, standing, or lying down. Keep your back straight.

Draw in a deep breath and count to *five* slowly. Exhale slowly and tell all your muscles to relax. Repeat this step two or three times until you are completely relaxed.

Imagine a pleasant thought, such as "I am learning how to relax," or a pleasant scene, such as a calm lake. If you use a natural scene, imagine all the sights, sounds, and smells of that scene as vividly as you can.

Practice this instant relaxation skill during your daily routine when you feel unwanted tension—for example, when you feel yourself becoming impatient while waiting in line.

Adapted from Farquhar, J. *The American Way of Life Need Not Be Hazardous to Your Health.* New York: W. W. Norton, 1978.

SESSION 4: TIME MANAGEMENT

Session Agenda

1. Preview the agenda for the session and make announcements.
2. Briefly review the content of Sessions 1, 2, and 3.
3. Practice the relaxation technique learned in Session 3.
4. Facilitate sharing of the Stress Management Action plans recorded in the Affirmation of Health forms.
5. Define "time urgency" and "time management."
6. Guide the group in identifying "time-wasters."
7. Provide a model for managing time.
8. Introduce "Twelve Mental Sets" useful in developing a more relaxed, positive approach to living.
9. Guide participants in filling out their Affirmation of Health forms.

*10. Evaluate the Stress Management course.

Handout

Handout 4A – Twelve Mental Sets: Suggested Attitudes to Enhance Wellbeing

Materials Needed

Flip chart
Magic markers
Session agenda printed on flip chart
Definitions of time urgency and time management (Activity 5) printed on flip chart
A model for time management (Activity 7) printed on flip chart
Course evaluation forms (available in Tips for Facilitators section)

Activity 1: PREVIEW THE AGENDA SESSION AND MAKE ANNOUNCEMENTS.

Methodology: Flip chart

Same as Activity 1 in Session 2 and 3.

---

*If this is *not* the final session of the course, Activity 10 should be altered to read "Evaluate the Session" and Activity 11 "Preview the Agenda for Session 5," should be added.

Activity 2: BRIEFLY REVIEW THE CONTENT OF SESSIONS 1, 2, AND 3.

Methodology: Flip chart
Group discussion

Ask the group to reflect upon the key ideas and skills they have learned. Record the list on the flip chart. Add any major concepts that are omitted.

Activity 3: PRACTICE THE RELAXATION TECHNIQUE LEARNED IN SESSION 3.

Methodology: Handout 3A, 3B, or 3C
Practice

Guide participants through the technique as in Session 3. Follow the instructions provided in handout.

Activity 4: FACILITATE SHARING THE STRESS MANAGEMENT ACTION PLAN RECORDED IN THE AFFIRMATION OF HEALTH FORMS.

Methodology: Handout 2E
Reportbacks

Same format as Session 3, Activity 3. The content of the sharing, however, will reflect the use of Session 3's relaxation technique in addressing stressful situations recorded in participants' Daily Stress and Tension Logs (Handout 2F) between Sessions 2 and 3.

Activity 5: DEFINE TIME URGENCY AND TIME MANAGEMENT.

Methodology: Group discussion
Lecturette
Flip chart

Time management is the stress management technique highlighted in this session. Ask participants how they would define time urgency and time management. Then provide the following two definitions:

Time urgency leads to stress in the individual. It is the feeling an individual has when he/she has taken on more tasks than he/she can accomplish in a given time. The individual is stressed both during the period of attempting to accomplish the tasks and later, when he/she is frustrated at not being able to get the work done.

Time management is an effective strategy to reduce the stress of time urgency. Time management involves allocating appropriate blocks of time for tasks and setting priorities for tasks to be completed.

This means maintaining a schedule to eliminate the need to worry about having insufficient time to carry out tasks. (Girdano & Everly, 1979, p. 150)

These definitions should be printed on the flip chart before class.

Activity 6: GUIDE THE GROUP IN IDENTIFYING TIME-WASTERS.

Methodology: Facilitator-led brainstorming

A. The whole group brainstorms a list of time-wasters in their lives.
B. The list is ranked in order of priorities. (This exercise also shows how priorities vary greatly among individuals.)
C. The list is divided to show which time-wasters are generated internally and which are generated externally. Each individual can then determine which items he/she can control. Internally generated time-wasters may require a change in values or attitudes.

Activity 7: PROVIDE A MODEL FOR MANAGING TIME.

Methodology: Flip chart
Lecturette
Small groups

Have the following model printed on the flip chart. Explain that it can be used for any given time period (i.e., a day, a week, a year).

A Model for Time Management

Time Demands	Time Supply
1. List all tasks to be completed in a given time interval.	1. Identify blocks of time available each day for completing necessary tasks.

2. Estimate time needed to complete each task.
3. Increase each time estimated by 10-15% (to allow for unexpected problems).

2. Match tasks with available time blocks.
3. Prioritize tasks so most important tasks are completed first.

Adapted from Girdano & Everly, 1979, p. 150.

Instruct the group to divide into pairs. Each person should discuss the application of this model to a specific day in the coming week. In other words, each will develop a plan for the management of his/her time for one day.

Activity 8: INTRODUCE TWELVE MENTAL SETS USEFUL IN DEVELOPING A MORE RELAXED, POSITIVE OUTLOOK ON LIVING.

Methodology: Reading aloud
Handout 4A

Refer participants to Handout 4A. Paraphrase the opening two paragraphs. Request individual participants to read aloud each of the twelve mental sets and the seven tips for using them. Encourage group discussion of these concepts, particularly concerning how the twelve mental sets might contribute to effective time management.

Activity 9: GUIDE PARTICIPANTS IN FILLING OUT THEIR AFFIRMATION OF HEALTH FORMS.

Methodology: Handout 2E

Allow the group to decide whether to work individually, with partners, or in small groups. Suggest that the goal-related activity for the coming week be using the model for time management on specified days, or practicing one or more of the twelve mental sets. If this is the final session, encourage participants to continue using the form after the course has ended. It is a tool for maintaining the changes they have made and for making more positive changes.

Activity 10: EVALUATE THE STRESS MANAGEMENT COURSE.

Methodology: Written evaluation form (available in Tips for Facilitators section)

If this is the final session of the course, distribute "Course Evaluation Forms" for all participants to complete in writing before leaving. Explain the importance of their honest, thorough evaluation in enabling the course to be continually improved for future participants.

(If this is not the final session, evaluate this session only, as was done in Sessions 1, 2, and 3; and review the agenda you have developed for Session 5.

HANDOUT 4A

TWELVE MENTAL SETS: SUGGESTED ATTITUDES TO ENHANCE WELLBEING

The twelve mental sets are self-statements for positive attitudes. Attitudes influence how we think, feel, and act. If you choose, you can practice the mental sets that are meaningful to you to help you increase a positive and relaxed approach to living. You can gradually weave them into your life fabric.

Please do not make these affirmations new "shoulds" for yourself. Be gentle with yourself. Results will become evident as you introduce these attitudes into your daily life.

Tips for using the twelve mental sets are included on the following page.

—I do the best I can about a situation, committing myself to its resolution without worrying.

—I set realistic goals for myself. When reasonable, I do one thing at a time.

—I am aware of my own feelings, and can choose to express them honestly to other people: I am responsible "to" others, not "for" them.

—I choose how to respond to stressful situations and accept responsibility for the consequences.

—I have no need to compare myself or to compete with other people.

—I treat all others with the respect and acceptance I wish for myself.

—I realize there are options in any given situation and I feel the freedom to explore.
—I learn lessons of growth from negative and positive experiences.
—I think and life positively, committing myself to achieving personal excellence; if I backslide, I can regroup and go on.
—Death is a normal, inevitable part of human life.
—I live in the present moment, realizing I can learn from the past and have hope for the future.
—By keeping in touch with my body and responding to its needs, I choose to be well and happy.

TIPS FOR USING THE TWELVE MENTAL SETS

It usually takes from one to four weeks of daily practice to gain some benefits from the mental sets. Regular and consistent practice is important. Don't worry about the results; they will come with time.

1. Choose the mental health sets which you believe will help you the most. Focus on those until you experience the results you want.
2. Revise the mental sets so they are relevant to your particular situation. For example, #3 can be changed to "I am aware of my own feelings, and express them honestly to my husband (neighbor, daughter, etc.)."
3. Pick a time of day for practice when you are relaxed and your mind is clear. For some people, early morning is best; for others, late evening or while walking in a park may be better.
4. If negative thoughts surface in your mind as you practice, let them pass without dwelling on them. It does take some time to change beliefs and attitudes, and in the meantime, old negative thoughts may appear. If they persist, you can say "stop!" then focus on the mental sets.
5. Repeat them *daily*, either silently or out loud, or write them down ten times daily. Spend five to ten minutes each day with them. Say or write them in a relaxed way, *as if* they were true for you (even if you are not quite convinced).
6. Imagine yourself living by the mental sets you have chosen.
7. Write the mental sets on 3″ × 5″ cards and place them somewhere (the refrigerator door, bathroom mirror, bedside table) where you will see and read them daily. Rewrite or move these cards to a new location every three or four days to reinforce noticing them.

SUGGESTED READINGS

Anderson, R. A. *Stress Power.* New York: Human Sciences Press, 1978.
Benson, H. *The Relaxation Response.* New York: Avon Books, 1976.
Berne, E. *Games People Play.* New York: Grove Press, 1964.
Blue Cross Association. *Stress*, U.S.A., 1974.
Bresler, D. E. *Free Yourself From Pain.* New York: Simon & Schuster, 1979.
Brown, B. *Stress and the Art of Biofeedback.* New York: Bantam Books, 1978.
Cooper, K. H. *The Aerobics Way.* New York: Bantam Books, 1977.
Cooper, K. H. *Aerobics.* New York: Evans & Co., 1968.
Cousins, N. *Anatomy of an Illness.* New York: W. W. Norton, 1979.
Devi, I. *Yoga for Americans.* Englewood Cliffs, NJ: Prentice-Hall, 1959.
Dyer, W. W. *Your Erroneous Zones.* New York: Funk & Wagnalls, 1976.
Farquhar, J. W. *The American Way of Life Need Not Be Hazardous to Your Health.* New York: W. W. Norton, 1978.
Friedman, M., & Roseman, R. H. *Type A Behavior and Your Heart.* New York: Fawcett Crest, 1974.
Girdano, D., & Everly, G. *Controlling Stress and Tension.* Englewood Cliffs, NJ: Prentice-Hall, 1979.
Goldwag, E. M. *Inner Balance—The Power of Holistic Healing.* Englewood Cliffs, NJ: Prentice-Hall, 1979.
Jacobson, E. *You Must Relax* (5th Ed.). New York: McGraw-Hill, 1978.
Keck, L. R. *The Spirit of Synergy: God's Power and You.* Nashville: Abingdon, 1978.
Lakein, A. *How to Get Control of Your Time and Your Life.* New York: The New American Library, 1973.
Leshane, L. *How to Meditate.* New York: Bantam Books, 1974.
Leshane, L. *You Can Fight for Your Life.* New York: Jove Publications, 1977.
Luce, G. G. *Your Second Life—Vitality and Growth in Middle and Later Years.* New York: Dell Publishing Co., 1979.
McCamy, J. C., & Presley, J. *Human Life Styling.* New York: Harper & Row, 1975.
Pelletier, K. *Mind as Healer, Mind as Slayer: A Holistic Approach to Preventing Stress Disorders.* New York: Delacorte and Delta, 1977.
Rosen, G. *The Relaxation Book: An Illustrated Self-Help Program.* Englewood Cliffs, NJ: Prentice-Hall, 1977.
Selye, H. *Stress Without Distress.* New York: J. B. Lippincott, 1974.
Shealy, C. N. *The Pain Game.* Millbrae, CA: Celestial Arts, 1976.
Shealy, C. N. *90 Days to Self-Health.* New York: Bantam Books, 1978.
Simonton, O. C., Matthews-Simonton, S., & Creighton, J. *Getting Well Again.* New York: St. Martin's Press, 1978.
Toffler, A. *Future Shock.* New York: Random House, 1970.
Walker, E. *Learn to Relax: 13 Ways to Reduce Tension.* Englewood Cliffs, NJ: Prentice-Hall, 1975.
White, J., & Fadiman, J. *Relax: How You Can Feel Better, Reduce Stress and Overcome Tension.* New York: Confucian Press, 1976.

Nutrition

Nutrition is a complex and challenging subject with new information (and controversies!) emerging all of the time. The four class sessions included in this section offer a range of topics and activities that can provide you and your participants with a beginning "taste" of Nutrition.

The class sessions as they are presented are crammed full of information and experiences. You will need to use them in a form which is compatible with the needs, interests, and skills of your participant group. Often, it will make sense to divide each of the four sessions presented here into several sessions. It is important to keep a balance of information presentation, skill development or practice, and discussion in each session, however long or short that session may be.

We recommend using guest speakers to help you deliver the content of these sessions. They bring knowledge, resources and a built-in "change of pace" to the program. For your nutrition class, some community resources that you could use include the dietitian, home economist or nutritionist from:

—County Agricultural Extension Office
—Public Health Department
—Hospital Outpatient Clinic
—Local Dietetic Association Office
—Local Heart Association Office
—Local Diabetes Association Office
—Local College or University Nutrition Department

Sources of Free Publications or Information

American Heart Association and its state associations have many free brochures with nutrition information. The Oregon Heart Association has a particularly good bulletin listing fat content of cheeses.

Blue Cross of Washington and Alaska, Public Relations Dept.,

P.O. Box 327, Seattle, WA 98111, provides recipes low in fat and cholesterol. It lists nutrients per serving for each recipe.

Consumer Information Center, Pueblo, CO 81009, is an excellent source of materials concerning nutrition, many of which are free or under $1.00. Recommended are: "Food-Information on Fat, Salt, Sugar, Fiber," "Ideas for Better Eating," "Dietary Guidelines for Americans," "Consumer's Guide to Food Labels," "Fats in Food and Diet," "Food Additives," "Grandma Called It Roughage," "A Primer on Dietary Minerals," "Salt," "Sugar."

Giant Foods, Inc., Consumer Affairs Department, P.O. Box 1804, Washington, D.C. 20013. A series of almanacs that provide nutritional information about fats, cholesterol, salt, and sugar. Aimed at the consumer. Very colorful and easy to follow.

Free information is generally available from state dairy councils. For a catalog of nutrition education materials write:

National Dairy Councils
6300 North River Road
Rosemont, IL 60018

The National Institution on Aging is involved in research on diet and the aging process. Write for general information or with specific questions to:

Information Office
National Institute on Aging
Building 31, Room 5C35
Bethesda, MD 20205
(301) 496-1752

Reprints of articles, audio-visuals, bibliographies and books are available through:

Food and Nutrition Information and
Education Resources Center
U.S. Department of Agriculture
Beltsville, MD 20705

Additional publications are listed at the end of this section.

Free Films

"Help Yourself to Better Health," AARP-Society for Nutrition Education

"Food, Energy and You," National Dairy Council

Rarig Film Service
834 Industry Drive
Seattle, WA 98188

"Nutrition from the Twenties through the Nineties"

Tupperware Home Parties
Educational Services
P.O. Box 2353
Orlando, FL 32802

Additional Ideas (for fun!!)

—Cooking demonstrations.
—Sampling of low fat, low sugar, or low salt refreshments, e.g., fresh, raw vegetables with a dip made with tofu instead of sour cream.
—Exchange of healthful recipes.
—Displays of food. This may be especially helpful with grains and legumes. Use labels that tell how to use these foods creatively.
—Create a bulletin board that associates good buys with healthy food. If broccoli (which is high in calcium) is on special, have a recipe for broccoli handy.

SESSION 1: AN OVERVIEW OF NUTRITION

Session Agenda

1. Welcome participants, preview the agenda, and make announcements.
2. Facilitate a discussion of participants' expectations of the nutrition course, explain the goals of the course, and familiarize the participants with course content.
3. Discuss the basic concepts of nutrition.
4. Discuss the special nutritional needs of older people.
5. Evaluate the session.
6. Preview the agenda for Session 2.

This plan assumes that the participants and the facilitator have met. If this is not the case, these activities should be preceded by introductions. Please refer to the Personal and Community Self-

Help component, Session 1, Activity 1, which describes an introductory game.

Handouts

Handout 1A – The Goals of the Class
Handout 1B – New American Eating Guide (or a substitute guide)
Handout 1C – Nutrients for Good Nutrition (or a substitute provided by guest speaker)
Handout 1D – Food Problems in Later Years (or a substitute provided by guest speaker)

Materials Needed

Flip chart
Magic markers
Agenda printed on flip chart
Agenda for Session 2 printed on flip chart

Activity 1: WELCOME THE PARTICIPANTS, PREVIEW THE AGENDA FOR THE SESSION, AND MAKE ANNOUNCEMENTS.

Methodology: Flip chart

After welcoming participants, direct their attention to the agenda printed on the flip chart. Read each item. Make any necessary announcements and invite participants to do likewise.

Activity 2: FACILITATE A DISCUSSION OF PARTICIPANTS' EXPECTATIONS OF THE NUTRITION COURSE, EXPLAIN THE GOALS OF THE COURSE, AND FAMILIARIZE PARTICIPANTS WITH COURSE CONTENT.

Methodology: Group discussion
Handout 1A

Ask each participant to share one thing he/she hopes to learn in the course or one change he/she would like to make in dietary habits.

Refer participants to Handout 1A. Paraphrase its contents or read it aloud.

Activity 3: DISCUSS THE BASIC CONCEPTS OF NUTRITION.

Methodology: Lecture
Group discussion
Handout 1B
Handout 1C

A guest speaker should be invited to discuss the basic concepts of nutrition. Following is an outline of the information to be presented.

1. The main nutrients and their functions:
   —carbohydrates (sugars and starches)
   —fats (animal and vegetable origins)
   —protein (animal and vegetable sources)
   —vitamins
   —minerals
   —roughage or fiber (a complex carbohydrate)
   —water (essential—although not a food, it is often classified as a nutrient)
2. Nutrients can be divided into four basic food groups:
   —bread and cereals (grains): 4 servings per day
   —fruit and vegetables: 4 servings per day
   —milk and milk products: 2 servings per day
   —meat and meat alternatives: 2 servings per day

   The "New American Eating Guide" (The Center for Science in the Public Interest) uses four groups, which are modified slightly from the above.
   —beans, grains, nuts: 4 or more servings per day
   —fruits and vegetables: 4 or more servings per day
   —milk products: 2 servings per day
   —poultry, fish, meat and eggs: 2 servings per day

   These groups are set into three classifications as a guide to better nutrition: Anytime, In Moderation, Now and Then. See Handout 1B
3. Energy sources:
   —starches and sugars
   —fats and oils (saturated and unsaturated)

4. Growth and maintenance:
   —complete proteins (animal origin)
   —incomplete proteins (vegetable origin)
   —complementary proteins
   —mixing and matching protein pairs
5. Protection:
   —minerals and vitamins

This presentation of information should be followed by a group discussion. It is suggested that handouts be provided. These may be supplied by the presenter or Handout 1C may be used.

Activity 4: DISCUSS THE SPECIAL NUTRITIONAL NEEDS OF OLDER PEOPLE.

Methodology: Lecture
Group discussion
Handout 1D

The same person who presented Activity 3 may present this information. Material to be included follows.

1. Deficiencies found in many "normal" older Americans:
   —iron deficiency (all income and ethnic groups)
   —vitamin A deficiency (mainly Spanish Americans)
   —vitamin C deficiency (more prevalent in men)
   —calcium deficiency (due to poor diet intake)
   —obesity (more women—high and low socioeconomic levels)
2. Fiber:
   —special importance for the elderly, particularly those who are inactive
   —supplementing the diet of those in institutions with high fiber foods
   —introduce fiber gradually into the diet when it has been lacking
3. Caloric needs:
   —decrease with age
   —need to have a well-balanced diet because of smaller intake of food than younger age groups (to ensure adequate vitamins, minerals and protein)

As in Activity 3, a group discussion should follow the presentation. Handout 1D may be used or the speaker may provide a handout with similar information.

Activity 5: EVALUATE THE SESSION.

Methodology: Group discussion
Flip chart

Explain that the purpose of the evaluation is to continually improve the course by incorporating participants' feedback and perceptions. To elicit comments, ask the group, "What were the most useful aspects of this session?" and "How could the session have been improved?" The facilitator should be receptive to constructive criticism.

Activity 6: PREVIEW THE AGENDA FOR SESSION 2.

Methodology: Flip chart

The agenda for Session 2 should already be written on the flip chart. Briefly read it aloud.

HANDOUT 1A

THE GOALS OF THE CLASS

Nutrition is an essential element of good health, and good health, in turn, insures a higher quality of life. Through gaining knowledge and accepting responsibility for your own health, you will be better able to make educated choices regarding your diet.

What we'd like to accomplish and how:

A. Provide useful information that will enable us to make educated choices regarding our diet. We will do this through:

1. class discussion
2. self-assessment tests
3. exploring healthful and tasty food
4. further reading and exploration of nutrition information.

B. Encourage positive changes in our eating habits, recognizing that changing our diet is a gradual, developmental process which can only be begun and supported in the time period of this program. We will do this through:

1. recording the foods we eat

2. using self-contracts
3. weekly sharing of progress made
4. cooking healthful food that tastes good
5. substituting healthier food for less healthy choices
6. modifying recipes to improve their nutritional value.

The *New American Eating Guide* which was prepared by the Center for Science in the Public Interest is considered controversial by some nutritionists. It is based on sound nutritional information, but differs in emphasis and style of presentation from the *Dietary Goals for the United States* (see reference bibliography). If a nutritionist is working with your program, you may wish to have her/him examine the *New American Eating Guide* and ascertain its appropriateness for your participant population. Otherwise, you may wish to use *Nutrition and Your Health: Dietary Guidelines for Americans,* these were prepared by the U.S. Department of Agriculture and published in 1981, if you have any doubts about using the *New American Eating Guide*.

HANDOUT 1B

NEW AMERICAN EATING GUIDE

Eating a healthy, balanced diet doesn't have to involve you in endless calorie counting or keep you forever tracking proteins, carbohydrates, fibers, cholesterol, or whatever. You can do it at a glance by using this simple chart put together by health and nutrition experts.

Plan your daily menus from each of the four groups paying special attention to these categories: anytime foods—these should be the backbone of your diet; *in moderation*—the next best category; *now and then*—eat these foods less often and in small portions.

When a certain food is placed in the "in moderation" or "now and then" column it means that there are certain disadvantages to over-indulging in that particular treat. For example, macaroni and cheese, (an "in moderation" food) uses a refined grain, contains moderate amounts of saturated fat, and may be high in salt. So, if you are concerned with reducing your intake of fats, salt, and sugar, this guide can help you make health-promoting decisions.

GROUP 1 - BEANS - GRAINS - NUTS

FOUR OR MORE SERVINGS PER DAY

ANYTIME:	IN MODERATION:	NOW AND THEN:
Bread and Rolls (whole grain)	Cornbread	Croissant
Dried Beans and Peas (legumes)	Flour Tortilla	Doughnut
Lentils	Granola Cereals	Presweetened Breakfast Cereals
Oatmeal	Macaroni and Cheese	Sticky Buns
Pasta, Whole Wheat	Matzo	Stuffing made with Butter
Brown Rice	Nuts	
Rye Bread	Pasta, Except Whole Wheat	
Sprouted Seeds	Peanut Butter	
Whole Grain Cereals	Pizza	
	Refried Beans, Commercial or Homemade in Oil	
	Seeds	
	Soybeans	
	Tofu	
	Waffles or Pancakes with Syrup	
	White Bread and Rolls	
	White Rice	

GROUP 2 - FRUITS - VEGETABLES

FOUR OR MORE SERVINGS PER DAY

ANYTIME:	IN MODERATION:	NOW AND THEN:
All Fruits and Vegetables (except those listed at right)	Avocado	Coconut
Applesauce (unsweetened)	Coleslaw	Pickles
Unsweetened Fruit Juice	Dried Fruit	
Unsalted Vegetable Juice	Fruits, Canned in Syrup	
Potatoes, White or Sweet	Glazed Carrots	
	Guacamole	
	Potatoes Au Gratin	
	Salted Vegetable Juices	
	Sweetened Fruit Juices	
	Vegetables, Canned with Salt	

GROUP 3 – MILK PRODUCTS

TWO SERVINGS PER DAY

ANYTIME:	IN MODERATION:	NOW AND THEN:
Buttermilk, Made From Skim Milk	Cocoa Made With Skim Milk	Cheesecake
Lassi (Low-Fat Yogurt and Fruit Juice Drink)	Cottage Cheese, Regular 4% Milkfat	Eggnog
Low-Fat Cottage Cheese	Frozen Low-Fat Yogurt	Whole Milk
Low-Fat Milk, 1% Milkfat	Ice Milk	Whole Milk Yogurt
Low-Fat Yogurt	Low-Fat Milk, 2% Milkfat	Hard Cheeses: Blue, Brick, Camembert, Cheddar, Muenster, Swiss
Nonfat Dry Milk	Low-Fat Yogurt, Sweetened	Ice Cream
Skim Milk Cheeses	Mozzarella Cheese, Part Skim Type Only	Processed Cheeses
Skim Milk		

GROUP 4 – POULTRY – FISH – MEAT – EGGS

TWO SERVINGS PER DAY

ANYTIME:

FISH
- Cod
- Flounder
- Gefilte Fish
- Haddock
- Halibut
- Perch
- Pollack
- Rockfish
- Sole
- Tuna, Water-Packed

EGG PRODUCTS
- Egg Whites Only

POULTRY
- Chicken or Turkey, Boiled, Baked, or Roasted (no skin)

IN MODERATION:

FISH (Drained Well, if Canned)
- Fried Fish
- Herring
- Mackerel, Canned
- Salmon, Pink, Canned
- Sardines
- Shrimp
- Tuna, Oil-Packed

POULTRY
- Chicken Liver, Baked or Broiled, just *one*!
- Fried Chicken, Homemade in Vegetable Oil
- Chicken or Turkey, Boiled, Baked, or Roasted (with skin)

NOW AND THEN:

POULTRY
- Fried Chicken, Commercially Prepared

EGG PRODUCTS
- Cheese Omelet
- Egg Yolk or Whole Egg (about 3/week)

RED MEATS
- Bacon
- Beef Liver, Fried
- Bologna
- Corned Beef
- Ham, Trimmed well
- Hotdogs
- Liverwurst
- Pigs' Feet
- Salami
- Sausage
- Spareribs

ANYTIME:	IN MODERATION:	NOW AND THEN:
	RED MEATS (Trimmed of all outside fat) Flank Steak Leg or Loin of Lamb Pork Shoulder or Loin, Lean Round Steak or Ground Round Rump Roast Sirloin Steak, Lean	
	VEAL	

Prepared by the Center for Science in the Public Interest, Washington, D.C., 1979.

HANDOUT 1C

NUTRIENTS FOR GOOD NUTRITION

Nutrient	Some reasons why you need it	Food Sources
	VITAMINS	
Vitamin A	Helps keep skin smooth and soft Helps keep mucus membranes resistant to infection Protects against night-blindness	Yellow fruits, green and yellow vegetables Butter, whole milk, cream, Cheddar-type cheese, ice cream Liver
Vitamin B_1 or Thiamine	Keeps appetite and digestion normal Keeps nervous system healthy Helps prevent irritability Helps body release energy from food	All meats, fish, and eggs Enriched and whole grain breads and cereals Milk White Potatoes
Vitamin B_2 or Riboflavin	Helps cells use oxygen Helps keep vision clear Helps prevent eyes from being unnaturally sensitive to light Helps prevent cracking at the corners of the mouth Helps keep skin and tongue smooth	Milk All kinds of cheese Ice cream Meats, fish, poultry Eggs

NUTRIENTS FOR GOOD NUTRITION (CONTINUED)

Nutrient	Some reasons why you need it	Food Sources
Vitamin B_6 or Pyridoxine	Important for health of blood vessels, blood cells and nervous system	Wheat germ, vegetables, meats, whole-grained cereals
B_{12}	Helps prevent certain forms of anemia Contributes to health of nervous system and to proper growth in children	Liver, kidney, milk, saltwater fish, oysters, lean meats
Folic Acid	Helps prevent certain forms of anemia Contributes to health, to nervous system, and to proper growth in children	Leafy green vegetables, meats
Niacin	Helps convert food to energy Aids nervous system and helps prevent loss of appetite	Lean meats, liver, enriched breads and cereals, eggs
Vitamin C (Ascorbic Acid)	Makes cementing materials that hold body cells together Makes walls of blood vessels firm Helps resist infection Helps prevent fatigue Helps in healing wounds and broken bones Important for healthy gums and teeth	Citrus fruits-lemons, orange, grapefruit, lime Fresh or frozen strawberries Cantaloupe Tomatoes Green peppers, broccoli Raw or lightly cooked greens or cabbage White potatoes
Vitamin D (the Sunshine Vitamin)	Helps the body absorb calcium from digestive tract Helps build calcium and phosphorus into bones	Vitamin D milk Butter Fish liver oil Sunshine (not a food)
Vitamin E	Acts as an antioxidant to preserve fat-soluble vitamins Believed necessary for reproduction	Vegetable oils, whole grain, cereals, wheat germ, lettuce
Vitamin K	Needed for the production of prothrombin which aids in the normal clotting of blood	Dark, leafy green vegetables

NUTRIENTS FOR GOOD NUTRITION (CONTINUED)

Nutrient	Some reasons why you need it	Food Sources
	MINERALS	
Calcium	Helps build bones, teeth Helps blood clot Helps muscles contract and relax normally Delays fatigue and helps tired muscles recover	Milk Cheese (but less in cottage cheese) Ice cream Turnip and mustard greens Collards and kale
Iron	Combines with protein to make hemoglobin, the red substance in the blood that carries oxygen to the cells	Liver Meat Eggs Green leafy vegetables Enriched breads and cereals
Iodine	Helps thyroid gland function properly by manufacturing the hormone thyroxine which regulates metabolism	Salt-water fish, shell-fish, iodized salt
Zinc	Affects growth, taste, healing, and reproduction	Found widely except in processed foods
	NUTRIENTS FOR ENERGY AND OTHER SPECIAL JOBS	
Protein (amino acids)	Builds and repairs all tissues in body Helps form substances in the blood called "antibodies" which fight infection Provides energy	Meat, fish, poultry All kinds of cheese Eggs Milk Cereals and breads Dried beans and peas Peanut butter
Fat	Supplies a large amount of energy in a small amount of food Helps keep skin smooth and healthy by supplying substances called "essential fatty acids"	Butter and cream Salad oils and dressings Cooking fats Fatty meats
Carbohydrates (sugars and starches)	Supplies energy Carries other nutrients present in the food	Breads and cereals Potatoes and corn Bananas, dried fruits, sweetened fruits Sugar, syrup, jelly, jams and honey

NUTRIENTS FOR GOOD NUTRITION (CONTINUED)

Water is also an essential nutrient, although we do not usually think of it as a food. Water is needed daily for:

Health of all the body cells
Help in carrying nutrients to cells and waste products away
Regulating body temperature
Aiding digestion
Replacing daily water loss
Building tissue

Source: Gutherie, H. A. *Introductory Nutrition.* St. Louis: C. V. Mosby, Co., 1971.

HANDOUT 1D

FOOD PROBLEMS IN LATER YEARS

The following chart is a summary of some common food and nutrition problems and suggested solutions:

NO APPETITE	—Avoid snacking, especially on sweet, caloric foods. —Don't worry, it is usual for taste buds to diminish, but a physician should check for other reasons. —Have regular mealtimes; eat the Basic 4. —Try walking or some other light exercise. —Start a hobby. —Acid foods are taste tinglers: fruit and juices, tomato juice, pickles, bouillon, dry wine.
LIMITED COOKING FACILITIES	—Nutritious meals can be uncooked. —Many meals can be cooked on a hot plate or electric skillet. —Make use of hearty one-dish meals such as stews and chowders with meat or fish and combined vegetables.
FOOD SPOILS EASILY WHEN COOKING FOR ONE	—Be sure to use leftovers effectively: label, freeze. —Buy smaller quantities—ask your butcher for the size you want. —Use recipes designed for one meal. —Cook cooperatively—get several friends together for meals, dividing costs and preparation.

CAN'T CHEW MANY FOODS; MOUTH HURTS	—Have dentures fitted properly. —Cube, chop, or grind foods difficult to chew. —Add broth to well-cooked, moist foods. —Use soft foods. Example: Meat loaf, omelettes, cottage cheese, blender-made vegetable soup.
EATING FRESH FRUITS AND VEGETABLES	—Buy in season—read grocery ads. Check farm bulk, a commodity exchange, or a senior food co-op. —Use frozen or canned fruit and vegetables if fresh are too expensive. —Combine with other foods, as in Oriental or Italian recipes.
MEAT IS EXPENSIVE	—Meat alternatives are just as nutritious. Use dried beans or peas, lentils, peanut butter, scrambled eggs, cottage cheese with soy bean curd. —Buy a larger cut, then divide into smaller cooking or serving portions.
FOOD BUDGET IS SMALL	—Food stamps stretch food dollars for those who qualify. —Eat at senior meal programs. —Take part in senior food co-ops.
OVERWEIGHT; DIABETES; HYPOGLYCEMIA (LOW BLOOD SUGAR)	—Use sparingly all kinds of fats and sweets. —Learn to say "NO" to second helpings. —Don't skimp on vegetables, grains, cereals, fresh fruits, skim milk, eggs, fish, and legumes. You need as much of these as ever. —Don't skip meals—plan to eat regularly, but less. —Start a mild, regular exercise program.
UNDERWEIGHT	—Eat concentrated foods, like cheese and thick chowders rather than broth and lettuce. —Add some snack foods such as eggnog, cocoa, ice cream, fresh or frozen fruits. —Gradually increase the amounts eaten. —Eat small amounts more often, including a complete protein each meal. (See "No Appetite" above.)
MUST AVOID SALT	—Avoid processed, packaged, and canned foods. Choose fresh or frozen. Avoid breaded meats and ham. —Get a magnifying glass. Read labels. Check for any form of "sodium."

—Be a gourmet cook with herbs, spices, lemon, onion, dill, fresh vegetables, dried vegetable flakes.
—Avoid diet pop, mixers, tube biscuits, dried soups.

Prepared by Kay Osborne, Nutritionist, King County Senior Services, Nutrition Programs, Seattle, Washington.

SESSION 2: BASIC STRATEGIES FOR DEVELOPING A NUTRITION PROGRAM

Session Agenda

1. Preview the agenda for the session and make announcements.
2. Briefly review the content of Session 1.
3. Discuss the health hazards in the average American diet.
4. Introduce and discuss an alternative to the average American diet.
5. Enable participants to increase their awareness of their dietary patterns.
6. Guide participants in establishing individualized nutrition improvement programs.
7. Evaluate the session.
8. Preview the agenda for Session 3.

Handouts

Handout 2A – Information Packet on Sugar, Fats, Salt, and Complex Carbohydrates
Handout 2B – Low-Fat Food Guide
Handout 2C – Benefits of Using the Seven-Day Diet Survey
Handout 2D – Seven-Day Diet Survey
Handout 2E – Daily Record of Cholesterol and Saturated Fat
Handout 2F – Daily Record of Salt, Sugar, Fiber, and Complex Carbohydrates
Handout 2G – Self-Contract Packet

Materials Needed

Flip chart
Magic markers
Agenda for the session printed on flip chart

Agenda for Session 3 printed on flip chart

Activity 1: PREVIEW THE AGENDA FOR THE SESSION AND MAKE ANNOUNCEMENTS.

Methodology: Flip chart

Read through the agenda items printed on the flip chart. After making announcements, invite participants to do likewise.

Activity 2: BRIEFLY REVIEW CONTENT OF SESSION 1.

Methodology: Flip chart

Request the group to briefly recap the main points covered in the presentations and discussions of Session 1.

Activity 3: DISCUSS THE HEALTH HAZARDS IN THE AVERAGE AMERICAN DIET.

Methodology: Lecture
Handout 2A

Facilitator or guest speaker presents introductory information on the ways in which the average American diet is hazardous to health. Major points to be included are:

1. Farquhar (1978) states that: "The average American diet is, in fact, decidedly hazardous to your health. It increases the incidence of cardiovascular disease and all the diseases associated with obesity (including diabetes, gout, osteoarthritis, gall bladder disease, and high blood pressure), and may well increase the incidence of certain types of cancer" (p. 103).
2. The American diet on the whole is too high in:
   —caloric density
   —saturated fat
   —cholesterol
   —salt
   —sugar

   The American average diet is too low in:
   —complex carbohydrates
   —fiber

Refer participants to Handout 2A. The information contained

in that handout should also be contained in the presentation. Allow time for questions following the presentation.

Activity 4: INTRODUCE AND DISCUSS AN ALTERNATIVE TO THE AVERAGE AMERICAN DIET.

Methodology: Lecturette
Group discussion
Handout 2B

Explain that the dietary guidelines for Americans established by the U.S. Dept. of Agriculture are congruent with the New American Eating Guide. These guidelines are:

—eat a variety of foods
—maintain ideal weight
—avoid too much saturated fat and cholesterol
—eat foods with adequate starch and fiber
—avoid too much sugar
—avoid too much sodium
—if you drink alcohol, do so in moderation

Source: U.S. Department of Agriculture, U.S. Department of Health, Education, and Welfare. *Nutrition and Your Health: Dietary Guidelines for Americans,* 1981.

Refer participants to Handout 2B. Ask one or two participants to read it aloud.

Activity 5: ENABLE PARTICIPANTS TO INCREASE THEIR AWARENESS OF THEIR DIETARY PATTERNS.

Methodology: Handout 2C
Handout 2D

Refer participants to Handouts 2C and 2D. Read Handout 2C aloud. The form is used to increase each person's awareness of his/her dietary patterns and dietary goals.

Request that everyone complete the form during the following week. Explain that these completed forms will be used in Session 3 to establish personal goals for the following week, and again in Session 4 when discussing "Eating for Maximum Nutrition."

Activity 6: GUIDE PARTICIPANTS IN ESTABLISHING INDIVIDUALIZED NUTRITION IMPROVEMENT PROGRAMS.

Methodology: Lecturette
Handout 2E
Handout 2F
Handout 2G

One tool to help make and maintain healthful changes is the personalized "self-contract." These enable participants to establish for themselves easily attainable short-term dietary goals.

In order to identify an appropriate goal for this week, participants will now examine their eating patterns individually through use of Handout 2E, the "Daily Record of Cholesterol and Saturated Fat" and Handout 2F, and the "Daily Record of Salt, Sugar, Fiber and Complex Carbohydrates."

Refer participants to these handouts. Read the introductory paragraph on Handout 2E, and ask participants to complete both 2E and 2F for the previous day.

Following this exercise, refer participants to Handout 2G which contains four self-contract forms, each focusing on a different food type. Instruct each participant to:

1. Using Handouts 2E and 2F, identify either salt, sugar, fats, or fibers and complex carbohydrates as the food type most in need of change in his/her diet; and
2. Choose the self-contract in Handout 2G which addresses that food type and complete it for the following week.

Remind participants that Session 3 will contain a section in which they may share with each other their experiences in seeking to fulfill the self-contracts.

Activity 7: EVALUATE THE SESSION.

Methodology: Group discussion
Flip chart

Same as Session 1, Activity 5.

Activity 8: PREVIEW THE AGENDA FOR SESSION 3.

Methodology: Flip chart

Same as Session 1, Activity 6.

HANDOUT 2A

INFORMATION PACKET ON SUGAR, FATS, SALT, AND COMPLEX CARBOHYDRATES

Sugar, Sugar Everywhere

Commonly eaten sugars and sweeteners do not provide any health benefits. They simply saturate the body with excessive calories. Natural sweeteners such as fruits and vegetables supply us with vitamins, minerals, and fibers as well as a sweet taste.

Sugar may be listed as sucrose, glucose, fructose, corn syrups, corn sweeteners, maltose, dextrose, invert sugar, and honey. Be especially aware of "new" cookbooks claiming sugar is not used. Many times these recipes rely almost 100% on honey. There are essentially no nutritional qualities distinguishing honey from white or brown sugar. Of course, an occasional cookie or teaspoon of sugar will cause no harm. *The important point is to learn to enjoy foods without added sweeteners and to condition your taste buds so that pre-sweetened foods will be too sweet.*

Fruits are often canned in heavy syrup, which is a high-sugar product. Buy fruit canned in its own juice or other fruit juice or water.

"Non-nutritive sweetener" on the label means an artificial sweetener has been added.

Bananas, raisins, and dried fruit can add sweeteners to many recipes. Now that they are often labeled as "health" foods, you must be careful to buy the ones without added sugar. Many supermarkets now feature these items with added sugar "for improved taste." Dried bananas do not need additional sugar.

Source: Williams, J., & Silverman, G. *No Salt, No Sugar, No Fat.* Concord, CA: Nitty Gritty Productions, 1981.

SOME EXAMPLES OF ADDED SUGAR IN PROCESSED FOODS

FOOD	SERVING SIZE	TEASPOONS SUGAR PER SERVING	% OF CALORIES THAT IS SUGAR
Cola Drink	12 oz.	9.2	99
Hi-C Orange Drink (Welch's Brand)	6 oz.	4.8	81
Kellogg's Fruit Loops	1 cup	3.5	49
Kellogg's Corn Flakes	1 cup	.5	7
General Mills Count Chocula	1 cup	3.2	46
Cranberry Sauce	½ cup	11.7	90
Catsup	1 tbsp.	.6	61
Yogurt, Fruit	8 oz.	7.5	50
Jello (cherry)	½ cup	4.5	87
Beets, pickled (Del Monte Brand)	½ cup	2.1	57

Source: Center for Science in the Public Interest, Washington, D.C.

Facts About Fat

All fats, saturated or polyunsaturated, provide *9 calories per gram* of fat. Everytime you use 1 tablespoon of vegetable oil, you consume 13.6 grams of fat or 122 calories.

Fats are classified as saturated or unsaturated depending on the type of fatty acid present. Saturated fats are fats that harden at room temperature, and they are found in most animal products and a few plant foods (coconut and palm oil). Polyunsaturated fats are usually oils and are most abundant in plants and fish oils. The most important polyunsaturated fat is linoleic acid. This is an essential nutrient that must be supplied by food because the body cannot manufacture it. The most recent *recommended dietary allowances* (RDA) suggests that we can meet this requirement if 1-2% of our calories are supplied by linoleic acid. Polyunsaturated vegetable oils are a good source of linoleic acid.

Only 30 percent or less of the calories you need each day should come from fat.* This is fat (oil) you can see in salad dressing, and fat you cannot see in cakes and pies, in a steak, and in an avocado.

Be aware of the surprisingly high fat content of most cheeses. Lowest in fat are dry cottage cheese, low-fat cottage cheese, and part-skim mozzarella. Highest are cream cheese and cheddar cheese. So many people substitute cheese for meat and think their diet has been improved. Unfortunately, the fat calories per serving in most whole milk cheeses equal or surpass beef. For example:

1 serving lean ground beef
(3 oz. cooked or 1 3 oz. patty) .... 9.6 grams of fat

1 serving cheddar cheese
(1 oz.) .... 9.1 grams of fat

1 serving part-skim mozzarella
(1 oz.) .... 4.7 grams of fat

Notice that the serving size of meat differs from that of cheese.

Stick to low-fat or nonfat dairy products. The difference between a cup of whole milk and a cup of skim milk is 72 calories of fat.

Keep in mind that prime grade beef contains more fat than choice grade beef, and choice more than good grade.

Nuts, peanuts, and peanut butter contain considerable amounts of fat. Chestnuts are the only nuts that contain small amounts of fat.

Fruits, vegetables, and grains contain very little fat and add important vitamins, minerals, and fiber.

Source: Williams, J., & Silverman, G. *No Salt, No Sugar, No Fat.* Concord, CA: Nitty Gritty Productions, 1981.

Please Don't Pass the Salt

Salt is sodium chloride. Sodium is the ingredient we are trying to avoid. Sodium bicarbonate, sodium benzoate, disodium phosphate, and mono-

---

*According to *Dietary Goals for the United States*, 2nd edition. Prepared by the staff of The Select Committee on Nutrition and Human Needs, United States Senate, 1977.

sodium glutamate are all names on labels indicating that sodium is present. Baking powder, baking soda, garlic salt, brine, or onion salt also contain sodium.

Recognize that salt is often an ''invisible'' ingredient in cakes, cereals, cheese, meat, catsup, mustard, canned vegetables, dried and canned soups, and many other foods that do not taste salty. For example, there are 236 mg of sodium in half a cup of canned asparagus while an equal amount of fresh, cooked asparagus contains only 1 mg.

Nutritional labels do not have to state how much sodium they contain. Even when salt is the last listed ingredient on the label, there is no way of knowing how much has been added.

Dietary goal concerning salt: approximately 1 tsp./day.

Source: Williams, J., & Silverman, G. *No Salt, No Sugar, No Fat.* Concord, CA: Nitty Gritty Productions, 1981.

Surprise!
Those Starchy Carbohydrate Foods You've Learned to Avoid Can Be Very Good for You

One of the *Dietary Guidelines* for America says: ''Eat foods with adequate starch and fiber.'' Here are three suggestions to get you started:

1. Eat a wide variety of carbohydrate foods such as vegetables, fruits, peas, beans, grains. Discover bulgur wheat and lentils!
2. Balance the decrease in fatty foods with an increase in carbohydrates—especially vegetables, fruits, and grains that provide fiber.
3. Starchy foods in moderate amounts may be part of a weight loss diet. It's the *extra* fats and calories you and the manufacturer add to starchy foods that hurt.

What do complex carbohydrates do?

—Provide nutrients that are processed out of refined foods.
—Provide energy more evenly for several hours.
—Provide bulk and roughage, aiding greatly in digestion and regularity.
—Provide a way to cut down on calories. An ounce of fat has *twice* the amount of calories as an ounce of carbohydrate.
—Provide a way to save you money. Meat and other foods high in animal protein take up a large portion of your food dollar. High carbohydrate foods can extend meat dishes. For example, serve a small

amount of chicken with vegetables over a lot of rice or use meat as a side dish rather than as the center of your meal.

Source: Williams, J., & Silverman, G. *No Salt, No Sugar, No Fat.* Concord, CA: Nitty Gritty Productions, 1981.

HANDOUT 2B

LOW-FAT FOOD GUIDE

If you've decided to reduce the amount of saturated fats, cholesterol, and calories in your food, here are some suggestions:

Foods to Try More Often

1. *Fish and poultry (without skin).* They're low in saturated fats, so try substituting them more often for meat.
2. *Lean cuts of meat.* When you eat red meat, choose lean cuts and trim visible fat. Broiling, boiling, roasting on a rack, or stewing meats helps remove fat, too.
3. *Low-fat dairy products.* Skim milk, low-fat milk, and nonfat dry milk offer a triple treat—less saturated fats, cholesterol, and calories.
4. *Fruits and vegetables.* They contain very little fat and *no* cholesterol! Easy to prepare, fruits and vegetables add variety in color, texture, and flavor to any meal.
5. *Grains and starchy foods.* Foods like enriched or whole grain breads, noodles, cereals, and potatoes are nutritious low-fat foods. They provide important vitamins, minerals, and fiber as well as help stretch your protein dollar. (Go lightly on high calorie or high fat "extras" like butter or sour cream.)
6. *Oils and margarines high in polyunsaturated fats.* Polyunsaturated fats are high in liquid vegetable oils—such as safflower, sunflower, corn, soybean, and sesame seed oils. Margarines that have twice as much polyunsaturated as saturated fats are also good sources of polyunsaturates. Polyunsaturated fats are thought to help lower blood cholesterol, but go easy on all fats. Their calories add up quickly.

Foods to Eat Less Often

1. *Foods High in Saturated Fats*

   A. *Fatty Meats.* Saturated fats are high in choice grades and cuts of beef, pork, ham, lamb, and fatty luncheon and variety meats.
   B. *High-fat dairy products.* Butter, cream, ice cream, whole milk, and cheese are high in saturated fats.
   C. *Fats and oils high in saturated fats.* Coconut and palm oil (used in many baked goods), many solid and hydrogenated* vegetable shortenings—such as butter, beef fat, and lard—are particularly high in saturated fats.

2. *Foods High In Cholesterol*

   A. *Organ meats.* Brain, kidney, and liver are high in cholesterol.
   B. *Egg yolks.* Are particularly high in cholesterol.
   C. *Shell fish.* Lobster, crab, oysters, and scallops are high in cholesterol.

HANDOUT 2C

THE SEVEN-DAY DIET PLAN

Tracking and recording dietary habits will help you become aware of what you eat. Conducting this survey at the start of a nutrition class provides you with a picture of what your diet looks like now. The seven-day diet plan can be used to plan a gradual, sensible, and personal program of dietary improvement. It will also serve as a measuring stick as you improve your diet.

Although many programs have found the seven day diet survey useful and effective, it may be difficult for some participants to keep such an extended and extensive record. Instead, have participants keep a twenty-four hour record one day per week for several weeks. Be sure to use a different day each week because any particular day may not present a realistic pic-

---

*Hydrogenated: Hydrogen is added to polyunsaturated liquid vegetable oils, generally to harden the fat. This may saturate the fat, but only the product's *label* or manufacturer can tell you the amount of polyunsaturated and saturated fats. If on a margarine label, the first or second ingredient is "hydrogenated," then the margarine is high in saturated fats. (The *intent* of the process of hydrogenation is not to lower the nutritive value of the hydrogenated product; rather it is to increase its shelf life.)

ture of the participant's dietary habits. One purpose of the diet survey is to help the participant develop a realistic idea of his/her current diet so that he/she can begin to work toward health-enhancing dietary changes which are desired. Any method which contributes to achieving this purpose is, of course, worth trying out.

If you continue to use it, the survey may also be an effective tool for altering eating habits. Keeping a daily record of foods consumed maintains your awareness of your dietary goals and actual eating habits.

Be sure to look for foods high in fat, sugar, or salt. The hidden nature of these items in foods often causes people to underestimate the amounts eaten. So, keep your pen and survey handy and discover if "you really are what you eat!"

HANDOUT 2D
SEVEN-DAY DIET SURVEY

	Breakfast	Lunch	Dinner	Snacks
1st day				
2nd day				
3rd day				
4th day				
5th day				
6th day				

HANDOUT 2D (CONTINUED)

	Breakfast	Lunch	Dinner	Snacks
7th day				

Instructions: Please record all food and beverages including seasonings and condiments you consume on each day. Estimate quantities (i.e., ½ tsp. salt, 1-5 oz. pork chop, 1 medium slice spice cake with frosting).

HANDOUT 2E

DAILY RECORD OF CHOLESTEROL AND SATURATED FAT

TO KNOW WHAT KINDS OF FOODS YOU ARE EATING, PLEASE COMPLETE THE FOLLOWING FORMS. TO DO THIS, MARK THOSE BOXES WHICH CONTAIN THE FOODS YOU EAT IN AN AVERAGE DAY. IT IS BEST TO DO THIS AFTER YOUR LAST MEAL OF THE DAY SO IT WILL INCLUDE ALL THE FOODS YOU ATE DURING THAT DAY.

CHOLESTEROL AND SATURATED FAT RECORD

FOOD TYPE	COLUMN 1	COLUMN 2	COLUMN 3	COLUMN 4
Milk	Whole milk	Some whole milk Some low-fat milk	Low-fat milk	Non-fat milk
Cheese	High-fat cheese Swiss, cheddar, Monterey Jack	Moderately low-fat cheese, tybo Fontina	Low-fat cheese Baker's cheese Farmer's cheese	Very low-fat cheese, cottage & Ricotta cheese
Cream	Whole cream, Half-n-half, ice cream	Powdered or liquid cream substitutes, ice milk	Sherbet	Cream substitutes without saturated or hydrogenated oils
Eggs	One or more eggs per day	One-half egg per day	One-quarter egg per day	None, egg whites only
Meats, Fish, Poultry, Liver	Liver, fatty red meats, duck or goose	Lean red meat, chicken, game hen or turkey with skin or deep-fried	Poultry without skin and not deep-fried, fish (not deep-fried)	Beans, Peas, grains, nuts, seeds
Spreads	Butter, lard, or margarine made with palm oil	Stick margarine	Soft tub margarine	No butter or margarine
Oils and Shortenings	Lard, coconut or palm hydrogenated vegetable oil	Hydrogenated oil, shortenings	Peanut or olive oil	Safflower, corn, cotton seed, sunflower or sesame oil
TOTALS:	_____ Points	_____ Points	_____ Points	_____ Points

Interpretation of Scores: Add up the number of boxes you marked in each column. The total is your score for that column. High scores in columns 3 & 4 indicate a diet low in cholesterol and saturated fats. High scores in columns 1 & 2 indicate a diet high in cholesterol and saturated fats.

From The Family Guide to Exercise and Nutrition, prepared by the Stanford Heart Disease Prevention Program.

HANDOUT 2F

DAILY RECORD OF SALT, SUGAR, FIBER, AND COMPLEX CARBOHYDRATES

FOOD TYPE	COLUMN 1	COLUMN 2	COLUMN 3	COLUMN 4
Salt	Ate usual amount table salt, Salts (onion, garlic, celery) soy sauce, steak sauce, catsup ☐	Used less salty seasonings than usual ☐	Used a salt substitute ☐	No salt added while cooking or eating ☐
	Ate usual amount of salty processed foods (sausage, ham, bacon, canned soup) ☐	Ate one-half of usual amount ☐	Ate one-fourth of usual amount ☐	Ate none ☐
	Ate usual amount of salted snack foods (salted nuts, potato chips, etc.) ☐	Ate less than usual amount ☐	☐	Ate none or ate unsalted varieties ☐
Sugar	Sweetened soft drinks - two or more cans per day ☐	One can per day ☐	One-half can per day ☐	None ☐
	Ate usual amount of candy, cake, pie, pastry, cookies ☐	Ate less than usual amount ☐	Ate only small amount ☐	Ate none ☐
	Canned fruits packed in heavy syrup ☐	Canned fruits in light syrup ☐	☐	Canned fruit in water ☐
	Ate usual amount of white sugar, jam, jelly, preserves, brown sugar, honey ☐	Ate one-half of usual amount ☐	Ate only a small amount ☐	None added while cooking or eating; bought no sugary foods ☐
Fiber and complex carbohydrates				
Breads, Grains, Cereals	White Bread, heavily refined cereals and grains (white rice, bleached flour, etc.) ☐		Lightly-milled breads, grains, cereals ☐	Whole-Grain breads & cereals (brown rice, whole-wheat bread, etc. ☐
Beans, Peas, Lentils			Canned ☐	Freshly cooked ☐
Other Vegetables			Canned or soft cooked ☐	Raw, fresh or crisply cooked ☐

SUBTOTAL: ____ Points ____ Points ____ Points ____ Points

HANDOUT 2F (CONTINUED)

DAILY RECORD OF SALT, SUGAR, FIBER AND COMPLEX CARBOHYDRATES

FOOD TYPE	COLUMN 1	COLUMN 2	COLUMN 3	COLUMN 4
Fruit		Juices without pulp ☐	Cooked or canned fruit, juices with pulp ☐	Fresh fruit ☐
SUBTOTAL:	____ Points	____ Points	____ Points	____ Points
SUBTOTAL FROM PREVIOUS PAGE:	____ Points	____ Points	____ Points	____ Points
GRAND TOTAL:	____ Points	____ Points	____ Points	____ Points

Interpretation of Scores: Add up the number of boxes you marked in each column. The total is your score for that column. High scores in columns 3 & 4 indicate a diet low in salt, sugar, and refined carbohydrates. High scores in columns 1 & 2 indicate a diet high in salt, sugar, and refined carbohydrates.

From The Family Guide to Exercise and Nutrition, prepared by the Stanford Heart Disease Prevention Program.

HANDOUT 2G

SELF-CONTRACT PACKET
SELF-CONTRACT FOR SUGAR

Name: ______________________ Date: __________

During the next week, I will try to accomplish the goal(s) checked below (choose only those goals you consider reasonable for yourself):

	Goal for This Week	Accomplished?
Sugar That Is Added to Foods		
—Reduce sugar added to coffee, tea, other drinks	______	______
—Reduce sugar put on fruit, cereal, and other foods	______	______
—Avoid artificial sweeteners	______	______
—Reduce sugar used in recipes	______	______

	Goal for This Week	Accomplished?
Sugary Foods		
—Cut down on pastry, cake, pie, cookies, candy, ice cream, ice milk, sherbet	______	______
—Buy foods containing less sugar	______	______
—Substitute unsweetened fruit juice with pulp for sweetened juices	______	______
—Substitute fruit for sweet desserts	______	______
—Substitute healthy snacks for sugary snacks	______	______
General		
—Try out a new and healthful recipe	______	______
—Make and share with (name: ____________) a new recipe	______	______
—Continue the good habits I have already incorporated into my lifestyle	______	______
—Additional goals:________ ______________________ ______________________	______	______

A week from today I will check in the second column the goal(s) I accomplished this week. If I reach my goal(s) for this week, I will reward myself in the following healthful way:____
____
____
____

SELF-CONTRACT FOR FATS AND CHOLESTEROL

Name: ______________________ Date: ___________

During the next week, I will try to accomplish the goal(s) checked below (choose only those goals you consider reasonable for yourself):

	Goal for This Week	Accomplished?
Butter, Margarine, Mayonnaise		
—Reduce/eliminate butter and hard margarine	_____	_____
—Reduce/eliminate mayonnaise	_____	_____
—Substitute soft tub margarine	_____	_____
—Substitute healthier spreads for mayonnaise	_____	_____
Lard, Shortening, Oil		
—Reduce/eliminate use of lard and shortening	_____	_____
—Reduce/eliminate coconut and palm oils	_____	_____
—Reduce/eliminate hydrogenated and partially hydrogenated oils	_____	_____
—Substitute polyunsaturated vegetable oil	_____	_____
Processed, Convenience, Baked Foods		
—Further reduce use of convenience foods, processed foods, commercially-baked foods, and fast foods that contain unhealthy fats	_____	_____
—Bake and cook foods myself as much as possible	_____	_____

	Goal for This Week	Accomplished?
Fast Foods		
—Reduce/eliminate deep-fried fast foods	______	______
Milk		
—Reduce/eliminate whole milk	______	______
—Mix milks to develop a taste for low-fat or nonfat milk	______	______
—Drink low-fat or nonfat milk	______	______
—Substitute low-fat yogurt for whole milk yogurt	______	______
Cream		
—Reduce/eliminate consumption of cream, half-and-half, and non-dairy creamers that contain coconut, palm, or hydrogenated oils	______	______
—Reduce/eliminate ice cream consumption	______	______
—Reduce/eliminate sour cream consumption	______	______
—Use healthy substitute for cream, sour cream, cream cheese, whipped cream, etc.	______	______
—Substitute sherbet or ice milk for ice cream	______	______

	Goal for This Week	Accomplished?
General		
—Continue the good habits I have already incorporated into my lifestyle	______	______
—Additional goals:________ ____________________	______	______

A week from today I will check in the second column the goal(s) I accomplished this week. If I reach my goal(s) for this week, I will reward myself in the following healthful way:____________________________
____
____

SELF-CONTRACT FOR COMPLEX CARBOHYDRATES

Name: ________________________ Date: ____________

During the next week, I will try to accomplish the goal(s) checked below (choose only those goals you consider reasonable for yourself):

	Goal for This Week	Accomplished?
Complex carbohydrates		
—Reduce consumption of heavily refined cereals and grains	______	______
—Reduce use of heavily refined cereals and grains in cooking	______	______
—Use foods made from whole grains and cereals	______	______
—Use whole grains and cereals in recipes	______	______

	Goal for This Week	Accomplished?
—Increase my consumption of beans, peas, lentils, starchy root vegetables, leafy vegetables, and fruits	______	______
—Substitute fresh fruits and vegetables	______	______
—Substitute fruit juices with pulp	______	______
General		
—Continue the good habits I have already incorporated into my lifestyle	______	______
—Additional goals:________	______	______

A week from today I will check in the second column the goal(s) I accomplished this week. If I reach my goal(s) for this week, I will reward myself in the following healthful way: ____________________________

____

____

SELF-CONTRACT FOR SALT

Name: ________________________ Date: ____________

During the next week, I will try to accomplish the goal(s) checked below (choose only those goals you consider reasonable for yourself):

	Goal for This Week	Accomplished?
Salt		
—Use less table salt on food	______	______
—Reduce salty flavorings and seasonings	______	______
—Cook with less salt	______	______

	Goal for This Week	Accomplished?
—Cook with fewer salty flavorings and seasonings	______	______
—Use seasonings and flavorings without salt	______	______
—Use salt substitutes	______	______
—Season food with herbs and spices instead of salt	______	______
—Reduce use of salty varieties of foods	______	______
—Reduce use of salty processed meats	______	______
—Read labels for salt content before buying foods	______	______
—Buy low-salt or no-salt varieties of foods	______	______
—Inquire about saltiness of foods at restaurants	______	______
—Order unsalty foods at restaurants	______	______
—Eat less fast foods that are salty	______	______
General		
—Continue the food habits I have already incorporated into my lifestyle	______	______
—Additional goals:________	______	______

A week from today I will check in the second column the goal(s) that I accomplished this week. If I reach my goal(s) for this week, I will reward myself in the following healthful way:______________________________

____

____

____

SESSION 3: PURCHASING FOOD FOR MAXIMUM NUTRITIONAL VALUE

Session Agenda

1. Preview the agenda for the session and make announcements.
2. Briefly review the content of Session 2.
3. Facilitate sharing of the use of the Seven-Day Diet Survey and the Self-Contracts.
4. Lead discussion of helpful tips on purchasing food for maximum nutritional value.
5. Develop participants' ability to interpret nutritional labeling on foods.
6. Define and discuss "junk foods."
7. Guide participants in further developing their personal nutrition improvement programs.
8. Evaluate the session.
9. Preview the agenda for Session 4.

Handouts

Handout 3A – A Shopping List for Health Substitutes
Handout 3B – The Good Cook's Guide for Nutritious Eating
Handout 3C – Read any Good Labels Lately???
Handout 2C – Seven-Day Diet Survey
Handout 2F – Self-Contract Packet

Materials Needed

Flip chart
Magic markers
Session agenda printed on flip chart
Agenda for Session 4 printed on flip chart
A few labels from food packaging (see Activity 5)

Activity 1: PREVIEW THE AGENDA FOR THE SESSION AND MAKE ANNOUNCEMENTS.

Methodology: Flip chart

Same as Session 2, Activity 1.

Activity 2: BRIEFLY REVIEW CONTENT OF SESSION 2.

Methodology: Group Discussion

Briefly go over the significant concepts covered in Session 2.

Activity 3: FACILITATE SHARING OF THE USE OF THE SEVEN-DAY DIET SURVEY AND THE SELF-CONTRACTS.

Methodology: Participants' Seven-Day Diet Surveys
Participants' completed Self-Contracts for previous week
Group discussion

Request participants refer to their completed Seven-Day Diet Surveys. Ask what their experiences had been in filling out the form (i.e., ''Was anyone surprised to realize what he/she was actually eating each day?'').

Request that participants refer to their Self-Contracts for the previous week. During a large group discussion, encourage individuals to describe their goals for the week, their endeavors in seeking to accomplish them, and the self-reward they used.

Recognize all endeavors regardless of success or failure. Emphasize that change takes time. Immediate results are not always to be expected; the very process of establishing and working toward goals is valuable.

Encourage participants to support each others' endeavors; facilitator's modeling of supportive responses is imperative. This peer support may also be developed by asking, in specific instances, how one might help another in accomplishing a goal. For example, one woman related that she had self-contracted to substitute a fresh fruit for a sugary dessert. She further related that although this was tasty enough, it did not meet her need to bake desserts—something she was very fond of doing regularly. Another participant might offer to assist her by sharing a number of recipes for baked desserts that were low-sugar and low-fat.

Activity 4: LEAD DISCUSSION OF HELPFUL TIPS ON PURCHASING FOOD FOR MAXIMUM NUTRITIONAL VALUE.

Methodology: Small Groups
Large group discussion
Handout 3A

Instruct the group to divide into small groups of four or five, where they will develop a list of tips on how to purchase food for maximum nutritional value. Provide them with the following guidelines to stimulate their discussion.

1. Where to shop
2. What to buy
   —choosing fruits and vegetables in season
   —choosing lean cuts of meat
3. How much to buy of various foods
4. Budgeting—how to eat well on a low budget by eating lower on the food chain (substituting grains for meat)
5. Use a shopping list—each small group develops a list
6. Beware of bargains such as stale vegetables with little nutritional value
7. Read labels
   —expiration date
   —net weight of contents
   —nutritional information
   —storage instructions

Reconvene the large group. Request that a spokesperson from each small group share the list generated. Encourage questions and clarifications among participants rather than between facilitator and participants. Explain that the small group lists will be collected and compiled, and that each participant will receive a copy of this list in Session 4.

Finally, refer participants to Handout 3A; ask one or two participants to read it aloud.

Activity 5: DEVELOP PARTICIPANTS' ABILITY TO INTERPRET NUTRITIONAL LABELING ON FOODS.

Methodology: Handout 3B
Handout 3C
Lecturette
Large group discussion

Refer participants to Handouts 3B and 3C. Briefly discuss the handouts. Present the participants with a few labels from food packages and ask them, as a group, to critique these. Finally, suggest that all participants begin critiquing labels while shopping, and that they bring a label to Session 4 to critique with the group.

Activity 6: DEFINE AND DISCUSS "JUNK FOODS."

Methodology: Group discussion

"Junk food" generally contains unhealthy amounts of salt, sugar and/or fat and not enough minerals, vitamins, protein or fiber.

Ask the group to list a number of junk foods; explain that ways of substituting more healthy foods will be discussed in Session 4.

Activity 7: GUIDE PARTICIPANTS IN DEVELOPING THEIR PERSONAL NUTRITION IMPROVEMENT PROGRAMS.

Methodology: Participants' Seven-Day Diet Surveys
Participants' Self-Contract Packets

Refer participants to their completed Seven-Day Diet Surveys (Handout 2D). Instruct them to circle those foods containing excessive amounts of salt, fat, or sugar.

Refer them to the "Self-Contract Packet" (Handout 2G) and instruct them to choose one of the four self-contract forms that will help them in reducing consumption of one of the frequently circled foods on the Seven-Day Diet Survey. Finally, ask that they fill out the contract for the following week.

Activity 8: EVALUATE THE SESSION.

Methodology: Group discussion
Flip chart

Same as the evaluation component in Sessions 1 and 2.

Activity 9: PREVIEW THE AGENDA FOR SESSION 4.

Methodology: Flip chart

Same as the final activity in Sessions 1 and 2.

HANDOUT 3A

A SHOPPING LIST FOR HEART-HEALTHY SUBSTITUTES

Instead of the Food You Usually Buy	Buy a Healthier Variety Instead	Or Better Yet, Substitute
beef, pork, lamb	very lean cuts veal	chicken, turkey fish, peas, lentils, grains, seeds, nuts, beans
whole milk	low-fat milk	nonfat milk
hard, whole milk cheeses (swiss, cheddar, monterey jack)	low-fat cheeses (mozzarella, farmer's)	very low-fat cheeses (low-fat cottage cheese, ricotta)
whole milk yogurt,	low-fat yogurt	
cream, half-and-half		cream substitutes without saturated fats
ice cream	ice milk	sherbet
butter lard	hard margarine peanut or olive oils	soft tub margarine safflower, corn soybean, sesame, or sunflower oil
salty foods	low salt varieties	unsalted varieties
cake, pies, cookies white bread heavily refined, low-fiber high-sugar breakfast cereals		fresh fruit whole-grain bread whole-grain, high fiber cereals
canned fruits and vegetables	frozen fruits and vegetables	fresh fruits and vegetables

Adapted from The Stanford Heart Disease Prevention Program's *The Family Guide to Exercise and Nutrition.*

HANDOUT 3B

IT'S ALL IN THE LABEL

Begin to read labels. Some of the information you need to know will be included there. If what you need to know is not stated, don't buy the product. Instead, write to the manufacturer that you are not buying the product because the label doesn't tell you enough. Specify what you want to know.

The ingredients in a product are listed on the label in order, by weight. In other words, *the ingredient that there is the most of is listed first*; the one there is the least of is listed last.

Nutrition information is given on a per-serving basis. The label indicates the size of a serving, the number of servings in the container, the number of calories per serving, and the amount in grams of protein, carbohydrate, and fat per serving.

The U.S. recommended daily allowances (U.S. RDAs) are the approximate amounts of protein, vitamins, and minerals that meet or exceed the needs of 80-90% of the adult population of this country. (Individual needs vary such that some people may require less.) Pregnant and lactating women and ill people comprise the remaining 10-20%.

"Natural flavorings" on the label can mean anything. For example, in a popular tomato sauce "natural flavorings" refers to oils derived from natural spices (red and green sweet peppers and red chili pepper).

Be especially careful when a label says "natural ingredients" or "only natural ingredients." Often these products are loaded with honey and/or vegetable oils.

"Low calorie food" contains no more than 40% calories per serving. "Reduced calorie food" must contain at least ⅓ fewer calories than a similar food that is not reduced, but must be at least equal nutritively to the food for which it is a substitute. "Substitute food product" may describe any simulated product whether it is made from a real or synthetic source.

Source: William, J., & Silverman, G. *No Salt, No Sugar, No Fat.* Concord, CA: Nitty Gritty Productions, 1981.

HANDOUT 3C

READ ANY GOOD LABELS LATELY?

When you are in the supermarket, let the nutrition label help you decide what to buy for your family. It can provide helpful information.

WHAT DOES A NUTRITION LABEL LOOK LIKE?

A nutrition label *must* tell you these things:

Nutrition information
(per serving) serving = 1 oz.
Servings per container = 12

Calories	110
Protein	2 grams
Carbohydrate	24 grams
Fat	0 grams

Percentage of U.S. recommended daily allowances (U.S. RDA)*

Protein	2
Thiamine	8
Niacin	2

*Contains less than 2 percent of U.S. RDA for vitamin A, vitamin C, riboflavin, calcium and iron

A nutrition label *may* include optional listings for types of fats, cholesterol, sodium and additional vitamins and minerals:

Nutrition information
(per serving) serving = 8 oz.
Servings per container = 1

Calories	560	Fat (percent of calories) 53%	13G
Protein	23 G	Poly-unsaturated	2G
Carbohydrate	43 G	saturated	9G
Sodium (365 mg/100 g)	840 MG		
Cholesterol* (20 MG/100 G)			40 MG

Percentage of U.S. recommended daily allowances (U.S. RDA)*

Protein	35	Riboflavin	15
Vitamin A	35	Niacin	25
Vitamin C	10	Calcium	2
Thiamine	15	Iron	2

*Information on fat and cholesterol content is provided for individuals who, on the advice of a physician, are modifying their total dietary intake of fat and cholesterol

"U.S. RDAs" are United States recommended daily allowances, the daily amounts of protein, vitamins and minerals for adults and children 4 or more years of age.

SESSION 4: EATING FOR MAXIMUM NUTRITION

Session Agenda

1. Preview the agenda for the session and make announcements.
2. Review the content of previous sessions with particular focus on Session 3.
3. Facilitate sharing of use of the self-contracts and the food purchasing tips.
4. Discuss methods of seasoning with less salt.
5. Discuss and demonstrate methods of cooking without fat.

6. Discuss methods of sweetening with less sugar.
7. Discuss and encourage the use of complementary vegetable proteins.

*8. Evaluate the nutrition course.

Handouts

Handout 4A – Season without Salt
Handout 4B – Hints for Cooking without Fat
Handout 4C – Hints to Sweeten without Sugar
Handout 4D – What about Protein?
Handout 4E – How to Mix and Match Protein Pairs
Handout 4F – Complementary Combinations

Materials Needed

Flip chart
Magic markers
Agenda for the session printed on flip chart
Course evaluation forms (found in "Tips for Facilitators")
The compiled list of "Helpful Tips on Purchasing Food for Maximum Nutritional Value" generated in Session 3, Activity 4 (a copy for each participant)
Equipment and ingredients for cooking demonstration (Activity 5)
Two food samples suggested in Activities 6 and 7

Activity 1: PREVIEW THE AGENDA FOR THE SESSION AND MAKE ANNOUNCEMENTS.

Methodology: Flip chart

Same as Activity 1 in Sessions 1 and 2.

Activity 2: REVIEW THE CONTENT OF PREVIOUS SESSIONS WITH PARTICULAR FOCUS ON SESSION 3.

Methodology: Copies of the list generated in Session 3, Activity 4

---

*If Session 4 is not the final session of the course, this activity will be changed and two additional activities added; specifications for so doing are found on the final page of this module.

Briefly list the major concepts covered in Sessions 1 and 2.

Pass out copies of the compiled list generated in Session 3, Activity 4 of "Helpful Tips on Purchasing Food for Maximum Nutritional Value." Remind participants that this list was created by them to use as an aid in food shopping.

Ask who brought in labels from food packages (as requested by facilitator in Session 3, Activity 5). As a group, critique two or three of these.

Activity 3: FACILITATE SHARING OF THE SELF-CONTRACTS AND THE FOOD PURCHASING TIPS.

Methodology: Group discussion

Same as Session 3, Activity 3 except, in this session, the sharing is on the use of the Self-Contracts and the food purchasing tips.

Activity 4: DISCUSS METHODS OF SEASONING WITH LESS SALT.

Methodology: Handout 4A
Group discussion

Refer participants to Handout 4A. Read it aloud or ask a participant to do so. Invite participants to suggest additional ways of reducing salt consumption.

Activity 5: DISCUSS AND DEMONSTRATE METHODS OF COOKING WITHOUT FAT.

Methodology: Handout 4B
Cooking demonstration

Refer participants to Handout 4B. Ask a different participant to read each segment aloud. Allow time for participants to discuss the handout. Invite them to suggest additional methods of cooking without fat.

Conduct a demonstration of cooking without fat if you have kitchen access. If not, bring in a prepared sample. Adhere to the following guidelines:

—The dish cooked should be one which typically contains fat.
—A written recipe should be passed out to each participant.

—Participants should be involved in the demonstration as much as possible.
—The dish prepared should be inexpensive, use easily obtainable ingredients, and be relatively simple to prepare.
—The demonstration should apply at least one of the hints provided in Handout 4B.

Sample the prepared food.

Activity 6: DISCUSS METHODS OF SWEETENING WITH LESS SUGAR.

Methodology: Handout 4C
Food sample
Group brainstorm
Flip chart

Refer participants to Handout 4C. Prepare one of these recipes in advance for sampling in class by participants.

Request participants to brainstorm additional examples of how to reduce sugar consumption. Record these on the flip chart.

Activity 7: DISCUSS AND ENCOURAGE THE USE OF COMPLEMENTARY VEGETABLE PROTEINS.

Methodology: Handout 4D
Handout 4E
Handout 4F
Group discussion
Food sample

Refer participants to Handouts 4D, 4E, and 4F. Paraphrase their contents to participants or read them aloud. Ask participants to provide other examples of complementary protein combinations. Again, it is suggested that a sample of one of the suggestions on Handout 4F be available for participants to taste.

Activity 8:* EVALUATE THE NUTRITION COURSE.

---

*If the nutrition course is to extend beyond four sessions, this evaluation activity should be changed to evaluation of Session 4, and should become Activity 9. Activity 8 should read: "Guide participants in developing individualized nutrition improvement programs;" the methodology for this activity would be the use of the Self-Contracts around issues discussed in Session 4. Activity 10 would then be to preview the agenda for Session 5.

Methodology: Group Discussion
Course evaluation forms (available in the "Tips for Facilitators" section)

Explain the importance of the participants' honest and thorough evaluations to continually improve the course for future participants. Pass out course evaluation forms and request that participants complete them before leaving.

HANDOUT 4A

SEASON WITHOUT SALT

Use Spices and Herbs!

Herbs and spices will enhance and enrich the natural taste of your food and eliminate the need for salt. Spices are defined as parts of plants, usually of tropical origin, such as the dried seeds, buds, fruit or flower parts, barks, or roots. Herbs are the leafy parts of temperate zone plants.

Spices and herbs should be stored in a cool, dry place in air-tight containers as they gradually lose flavor and color during storage.

There is no general rule for the amount of herbs or spices to use. The pungency of each differs and its effect on different foods varies. Generally, if a recipe is not available, start with ¼ teaspoon and increase as directed.

The flavor of ground spices is imparted immediately; they may be added about 15 minutes before the end of cooking. Whole spices are best added at the beginning so long simmering can extract the full flavor and aroma. Whole or leaf herbs should be crumbled finely just before they are used to release the flavor. The flavor of herbs dwindles with long cooking.

If a recipe calls for fresh herbs, substitute 1 teaspoon dried for 1 tablespoon fresh.

Easy ways to beat those salty snack attacks

—Remove salt shakers from the kitchen and dining room. If it's not there, you can't use it.
—Buy or prepare snacks low in sodium, such as fruits, crisp vegetables, and unsalted nuts. If high sodium snacks aren't in the house, they'll be hard to eat.
—When preparing snacks, season foods with spices such as basil, dill, and lemon juice.

HANDOUT 4B

HINTS FOR COOKING WITHOUT FAT

Brown Without Fat

1. Use the broiler.
   Preheat broiler. Place food to be broiled on rack in broiler pan. Heat until brown, turn to brown all sides. Use this method for meat, poultry, fish, and vegetables. It's fast and simple, and your range top stays clean.
2. Use a wok or heavy iron skillet.
   Cut meat into bite-sized pieces. Heat empty pan, add meat, and stir-fry until brown. Remove fat with baster or spoon. Grains may be browned in this manner, but remember to stir constantly to prevent burning.

Sauté Without Shortening

Place ¼ to ½ cup of stock (chicken or vegetable) or water in a skillet or pan. Add a few herbs. Heat until simmering, add ingredients (usually chopped vegetables), and heat until vegetables are tender and slightly brown.

This method takes 2-10 minutes, depending upon size and kind of vegetable. Stir frequently to keep from burning.

You can stir-fry in a nonstick pan and omit the liquid, but the stock imparts flavor to the vegetables.

Marinate Without Oil

By changing herbs, spices, and liquid base, you can completely change the taste of a dish and enhance its flavor. Recipes will give specific amounts, but here are some general ideas.

1. Use yogurt. Yogurt and juices (lemon, orange, and tomato) combined with herbs and spices make a tasty marinade for chicken, beef, fish, vegetables, and grains.
2. Use sauces. Tomato sauce, tomato paste, tomato juice, Mexican sauces, and wine combined with onions and garlic are excellent for fish, chicken, and vegetables.

NOW THAT YOU HAVE SOME FACTS, HERE ARE HEALTHIER WAYS TO COOK YOUR FOOD

Meat

—Cook meat in cookware that enables you to discard the melted fat.
—Roast, barbeque, broil, pan-broil, braise, or stew meat to remove fat.
—Trim fat off meat before cooking it.

Dairy products

—*Butter:* substitute margarine (choose one which lists a liquid, nonhydrogenated oil as its first ingredient).
—*Cream:* substitute evaporated skim milk.
—*Sour cream:* substitute ¾ cup washed, large curd low-fat cottage cheese with 1-2 tablespoons milk or buttermilk. Whip together. Or, whip ¾ cup low-fat cottage cheese and 1 tablespoon lemon juice.
—*Whipped topping:* substitute ¼ cup nonfat milk powder sprinkled with ¼ cup ice water. Beat until thick. Then add ¼ teaspoon vanilla, ½ teaspoon lemon juice, and sugar to taste.
—*Whole milk:* substitute low-fat milk, nonfat milk, or buttermilk.
—*Cream cheese:* substitute neufchatel cheese or washed low-fat cottage cheese.
—*Packaged cake mixes:* they usually contain a saturated fat, such as coconut oil, palm oil, lard, or hydrogenated vegetable oil. Bake cakes from scratch!

Eggs

—*One whole egg:* substitute 2 egg whites and 1 teaspoon polyunsaturated vegetable oil.

Source: Williams, J., & Silverman, G. *No Salt, No Sugar, No Fat.* Concord, CA: Nitty Gritty Productions, 1981.

HANDOUT 4C

HINTS TO SWEETEN WITHOUT SUGAR

Reduce sugar or honey by half (or more) in all cookie recipes.

Use cinnamon, vanilla, raisins, mashed bananas, grated apple, chopped dates, or concentrated fruit juices as sweeteners in your favorite recipes.

Serve unsweetened fruit juices on ice instead of powdered drink mixes.

Sweet but healthy drinks:

FRUIT SHAKE

½ c. nonfat milk
2 T. instant nonfat dry milk
2 ripe bananas, pears, or peaches
¼ t. cinnamon
1 T. apple juice concentrate

Place all ingredients in blender and blend for 30-45 seconds. Serve cold.

FRUIT GELATIN

1 c. water
1 c. orange juice
1 package unflavored gelatin softened in cold water (see instructions below)
1 c. pineapple tidbits

Combine all ingredients except the pineapple in a saucepan and bring to a boil while stirring. Cool. Pour into a dish and add the pineapple. Chill until firm.

GELATIN

Make gelatin with unflavored gelatin and fruit juices. If you must add a teaspoon of sugar, go ahead. You are still ahead of commercial jello which contains about 5 teaspoons of sugar per ½ c. serving.

PANCAKE AND WAFFLE SAUCE

Sweeten your pancakes with your own sweet sauce by making it yourself. *You* control the amount of sugar used.

FRUITY SAUCE

1 c. unsweetened fruit juice
1 T. arrowroot or cornstarch

Dissolve arrowroot or cornstach in an equal amount of water before adding it to fruit juice. Heat over medium heat. Stir until thickened. Serve warm or cool. This sauce keeps in the refrigerator five days. Keep covered.

HANDOUT 4D

WHAT ABOUT PROTEIN?

The proteins in your body are made up of 22 different amino acids. Your body can make 14 of these acids. It cannot make 8 of them. These are the 8 "essential" amino acids, which must be supplied by the foods we eat.

Some foods have "complete proteins"—they contain the 8 essential amino acids in the right amounts. All foods from animal sources such as milk, poultry, fish, eggs, cheese, and meat, have complete protein. (Gelatin is the one exception—it is the only incomplete protein from an animal source.) However, many of these foods contain high amounts of saturated fats, cholesterol, and calories.

So, begin to learn about protein from vegetable sources. Legumes, such as beans, grains, nuts, and seeds, are a good source of protein and are real heart pleasers—low in saturated fats, no cholesterol—high in variety, color, and fiber.

A few facts. Vegetable sources of protein—like grains, dried beans, peas, nuts, and seeds—are usually low in one or another of the 8 amino acids and form an incomplete protein. These "incompletes" become "completes" when paired with other incomplete protein foods at the same meal. By adding another protein source to complement the essential amino acid which is low in a particular protein food, it is possible to prepare meat substitutes equal to animal products in terms of protein quality.

Protein Complementary Chart (heavy lines show best ways to complement proteins, e.g., a dish containing rice and beans is high in protein) from *The Family Guide to Exercise and Nutrition: Step-by-Step to a Healthy Heart*, prepared by the Stanford Heart Disease Prevention Program.

HANDOUT 4E

HOW TO MIX AND MATCH PROTEIN PAIRS

To make protein rich combinations, you can: 1. match vegetable proteins: mix foods from two or more groups in column A, such as peanut butter (a legume product) + whole wheat bread (a grain product), 2. match vegetable and low-fat animal products: mix foods from any group(s) in column A with small amount from any group(s) in column B, such as rice (a grain) + chicken (a low-fat meat).

COLUMN A
VEGETABLE PROTEINS
(INCOMPLETE)

Legumes

Dry beans and peas—kidney, navy, lima, pinto, black, or soy beans, black-eyed or split peas, soy bean curd (tofu), soy flour, peanuts and peanut butter (use sparingly: has medium fat level)

Grains

Whole grains—barley, oats, rice, rye, wheat (bulgur, cracked wheat), corn; pasta—noodles, spaghetti, macaroni, lasagna

COLUMN B
LOW-FAT
DAIRY PRODUCTS

Nonfat dry milk,
skim milk,
low-fat cottage cheese,
egg whites (where most of the egg protein lies)

low-fat meats—poultry, fish, lean cuts of red meat

Nuts and Seeds

Almonds, cashews, pecans and walnuts, sunflower seeds, pumpkin seeds, sesame seeds

Source: "Eater's Almanac," Consumer Affairs Department, Giant Food Inc., P.O. Box 1804, Washington, D.C. 20013.

HANDOUT 4F

COMPLEMENTARY COMBINATIONS

Here are some suggestions for combining foods to complement the protein in the foods. There are many more possible combinations. Use your imagination!

Breakfast

—granola cereal (oats, wheat, nuts, seeds) with nonfat milk
—bran cereals with nonfat milk and fruit
—whole wheat toast with peanut butter and raisins
—plain low-fat yogurt with wheat germ and fresh fruit
—fruit "smoothie" made with low-fat yogurt, nonfat milk, fresh fruit, bran, cinnamon, mixed in a blender

Lunch

—chili beans with cornbread
—navy bean soup and whole wheat bread
—low-fat yogurt with granola and fresh fruit
—tostadas with beans, lettuce, tomato avocado, onion, low-fat cheese, and plain low-fat yogurt
—turkey, chicken, or tuna fish on whole wheat bread

Dinners

—whole wheat pasta with meatless sauce
—beans and rice

—zucchini stuffed with rice and peas
—stir-fried or steamed vegetables with tofu, sunflower seeds, sesame seeds and rice
—spinach lasagne with ricotta cheese and mushrooms

Desserts and Snacks

—banana, zucchini, or poppy seed bread
—rice and milk pudding
—fresh fruit salad with low-fat yogurt and cinnamon

Adapted from the Stanford Heart Disease Prevention Program.

SUGGESTED READINGS

The references in the "Cookbook Bibliography" are included as sources of recipes, healthy cooking techniques, food substitutions and food combinations. Their inclusion in this bibliography should not be taken as a recommendation of any particular dietary plan advocated in the cookbooks.

COOKBOOK BIBLIOGRAPHY

Laurel's Kitchen, Laurel Robertson, Carol Flinders, and Bronwen Godfrey. Bantam Books, 1978. Over 400 recipes for vegetarian cooking. Directions for cooking with grains, legumes, and unusual vegetables. Good discussion of nutrients and easy to understand charts on vitamins, minerals, fat, calories, etc. One of the best vegetarian cookbooks. The authors are very careful to use small amounts of fat and sugar.

Live Longer Now, Jon Leonard. Grosset & Dunlap, 1977. Excellent collection of low fat, low sugar, low salt recipes. Includes shopping guide and discussion of why this type of cooking is beneficial. Excellent glossary.

Pritikin Program for Diet and Exercise, Nathan Pritikin with Patrick N. McGrady, Jr., Grosset & Dunlap, 1979. Everything you ever wanted to know about the Pritikin Diet. Good collection of recipes. Pritikin uses no fat, salt, or sugar. Directions are easy to follow. Uses lots of spices, grains, and vegetables. Includes list of acceptable foods. Includes recipes for the single cook. Based on a program designed for people who have suffered heart disease and stroke.

Light Style, Rose Dosti, Deborah Kidushim, and Mark Wolke. Harper & Row, New York, 1979. Good collection of low fat, low salt, low sugar recipes. Includes amounts of cholesterol, fat, and sodium for each recipe. Gourmet recipes for special occasions. Ways to make egg substitutes and low sodium soy sauce.

Pure and Simple: Delicious Recipes for Additive Free Cooking, Marion Burros. Berkeley Books, 1978.

Mushrooms and Bean Sprouts, Norma M. MacRae, R.D. Pacific Search, 1979. All recipes use mushrooms and sprouts which complement each other for better use of protein. Includes substitutions for low-calorie and low-cholesterol diets.

The American Heart Association Cookbook (3rd edition), McKay, NY, 1977. Low fat, low cholesterol, low sodium recipes. Available in paperback.

The Alternative Diet Book, William E. Connor, Sonja L. Connor, Martha M. Fry, and Susan L. Warner. The University of Iowa, 1976. Good collection of recipes for those interested in reducing fat and cholesterol. Gives information on a gradual approach to improving the diet.

Diet for a Small Planet, Frances Moore Lappe. Ballatine Books, 1971. Discusses foods that complement each other to make delicious protein rich meals without heavy use of meat. Discussion of amino acids and how much protein we need. Good recipes although some may need adapting to cut amounts of fat, salt, and sugar.

Recipes for a Small Planet, Ellen Buchman Ewald. Ballatine Books, 1975. A collection of high-protein vegetarian recipes using natural foods. Recipes may have to be adapted to reduce amounts of fat, salt, and sugar. Recipes show unusual ways to use grains, legumes, and vegetables. The presentation of protein nutrition could be simplified.

The Wonderful World of Natural-Food Cookery, Eleanor Levitt. Hearthside Press, Inc., 1971. Answers all your questions about health foods. Provides information on buying and cooking fruits and vegetables. Chapter on grains is especially good. Recipes can be adapted to reduce amounts of fat, salt, and eggs.

REFERENCE BIBLIOGRAPHY

The Dictionary of Sodium, Fats and Cholesterol, Barbara Kraus. Grosset & Dunlap, 1974. Fat, sodium, and cholesterol levels are listed for over 9,000 common foods. Includes information on natural and prepared foods.

Nutritive Value of American Foods in Common Units, Agriculture Handbook No. 456, U.S. Department of Agriculture. Washington Government Printing Office, 1975. (This is now being updated.) Lists values for nutrients in common foods.

Recommended Dietary Allowances, 9th Edition. National Academy of Sciences. Washington, D.C., 1980. A listing of the essential nutrients established for healthy populations. Charts indicate caloric needs as well as recommended protein, carbohydrate, fiber, vitamin, and mineral intake.

Dietary Goals for the United States, 2nd Edition. Prepared by the staff of The Select Committee on Nutrition and Human Needs, United States Senate, 1977. Practical guide for good nutrition. Sets standards for amounts of fat, carbohydrates, cholesterol, and sodium considered necessary for healthful diet. Recommends way for government to implement these guidelines.

The Supermarket Handbook, Nikke Goldbeck and David Goldbeck. New American Library, New York, NY: 1976. Teaches the reader to examine labels and recognize what specific ingredients should or should not be included in the products we purchase at the supermarket. Includes recipes for making your own soups and sauces.

Nutrition: Concepts and Controversies. E. V. Hamilton and E. N. Whitney. West, 1979. Presents a good discussion of controversies in the field of nutrition.

The Family Guide to Exercise and Nutrition. Stanford Heart Disease Prevention Program. A valuable resource for developing educational materials. Presents a week-by-week program for changing dietary and exercise habits.

ADDITIONAL READINGS

The books included in the "Additional Readings" section represent a variety of perspectives on nutrition. Some of them would not be considered "mainstream dietary information" by the American Dietary Association and therefore commonsense, good judgment and the advice of a local nutritionist is appropriate when developing your reading list for participants.

Nutrition is a controversial subject. We encourage you to read these materials with a critical eye, testing their information against class materials, scientific nutrition texts, common sense, and your own personal experiences.

Cleave, T. L. *The Saccharine Disease.* CT: Keats Publishing, 1975.
Cross, Jennifer. *The Supermarket Trap.* IND: Indiana University Press, 1976.
Davis, Adele. *Let's Eat Right to Keep Fit.* NY: A Signet Book, 1970.
Deutsch, R., *Realities of Nutrition.* NY: Bell Publishers, 1976.
Farquhar, John. *The American Way of Life Need Not Be Hazardous to Your Health.* NY: W. W. Norton, 1978.
Jacobson, Michael. *Nutrition Scoreboard.* NY: Avon Books, 1975.
Lerza, C. & Jacobson, M. (Eds). *Food for People, Not for Profit.* NY: Ballantine Books, 1975.
Mayer, Jean. *Human Nutrition: Its Psychological, Medical and Social Aspects.* NY: C. C. Thomas, 1979.
Mayer, Jean. *A Diet for Living.* NY: Pocket Books, 1975.
McCamy, John. *Human Life Styling.* NY: Harper & Row, 1975.
National Institute on Aging, "Food: Staying Healthy After 65," *Age Page*, Dec. 1980, Bethesda, MD 20205.
Nutrition Almanac, from Nutrition Search, Inc. NY: McGraw-Hill, 1975.
Price, Weston. *Nutrition and Physical Degeneration.* CA; Pothinger Foundation, 1965.
Robbins, Wm. *American Food Scandal.* NJ: Wm. Morrow & Co., 1974.
Rockstein, Morris (Ed.) *Nutrition, Longevity and Aging.* NY: Academic Press, Inc., 1976.
Watson, George. *Nutrition and Your Mind: The Psychochemical Response.* NY: Harper & Row, 1972.
Williams, R., M.D. *Nutrition Against Disease.* NY: Bantam, 1973.
Yudkin, J., M.D. *Sweet and Dangerous.* NY: Peter Syden, 1973.

The National Clearinghouse on Aging, SCAN Social Gerontology Resource Center (P.O. Box 231, Silver Spring, MD 20907 (301) 565-4269), publishes a "Nutrition/Nutrition Programs Bibliography" which contains more than thirty references on nutrition and aging, including technical, scientific, and programmatic books and articles.

Physical Fitness

The four class sessions presented here will provide you with some general information about physical fitness for older persons and a few exercise routines that can be adapted for use with most populations.

For assistance in planning activities, delivering information or further developing your exercise program consider calling upon guest speakers and co-facilitators from:

—YMCA/YWCA
—American Heart Association
—Arthritis Foundation
—City or County Parks and Recreation Department
—Community Centers
—Senior Centers
—Hospital-based Physical Therapy Departments

Helpful Publications

A National Directory of Physical Fitness Programs for Older Adults lists colleges and universities that offer innovative fitness programs for the well and disabled elderly. Information on individual programs, locations, activities and target populations are offered. Available for $4.00 from North County Community College Press, Saranac Lake, NY 12983.

An Older Americans Physical Fitness Program is available for $2.00 prepaid from the University Gerontology Center, Wichita State University, Box 121, Wichita, KS 67208.

Basic Exercises for People Over Sixty and Moderate Exercises for People Over Sixty available for $1.00 donation each from the National Association for Human Development, 1750 Pennsylvania Avenue, N.W., Washington, DC 20006.

Exercise and Your Heart (NIH Publication No. 81-1677) available through U.S. Department of Health and Human Services, Public Health Service, National Institute of Health.

Pep Up Your Life: A Fitness Book for Seniors and *The Good Life: A Physical Fitness Program for Senior Citizens* are available from the Travelers Insurance Companies, One Tower Square, Hartford, CT 06113.

60+ and Physically Fit: Suggested Exercises for Older People. State of Connecticut, Department on Aging, Physical Fitness Committee, 90 Washington Street, Hartford, CT 06113.

"Walking . . . The Preferred Exercise for Seniors" (pamphlet) is made available by the Massachusetts Department of Elder Affairs, 110 Tremont Street, Boston, MA 02108.

Additional publications are listed at the end of this section.

SESSION 1: PHYSICAL FITNESS FOR OLDER ADULTS—AN OVERVIEW

Session Agenda

1. Preview the agenda for the session and make announcements.
2. Facilitate discussion of participants' expectations of the course and the goals of the course.
3. Familiarize participants with an overview of the course.
4. Distinguish between the two main types of exercise and facilitate discussion of the benefits of regular exercise, with emphasis on the elderly.
5. Introduce and practice Stretch Break Movements.
6. Describe five normal responses to exercise and seven responses indicating inappropriate exercise.
7. Discuss situations in which caution in exercise is advisable, and pass out the "Release on Liability Statement and the "Personal Release of Liability Statement." (Optional)
8. Explain the Daily Physical Activity Log and encourage its use.
9. Evaluate the session.
10. Preview the agenda for Session 2.

NOTE: This plan assumes that participants and facilitator have met. If this is not the case, these activities should be preceded by introductions. Refer to the Personal and Community Self-Help component, Session 1, Activity 1, for an introductory game.

Handouts

Handout 1A – The Goals of the Class
Handout 1B – Physical Fitness Course Overview
Handout 1C – Benefits of Regular Exercise
Handout 1D – Stretch Break Movements
Handout 1E – Responses to Exercise
Handout 1F – Daily Physical Activity Log

Materials Needed

Flip chart
Magic markers
Session agenda printed on flip chart
Agenda for Session 2 printed on flip chart
Release of Liability Statements
Personal Release of Liability Statements

Activity 1: PREVIEW THE AGENDA FOR THE SESSION, AND MAKE ANNOUNCEMENTS.

Methodology: Brief welcoming talk
Flip chart

After welcoming participants, direct their attention to the agenda written on the flip chart. Briefly describe each item. Make any necessary announcements and invite participants to do likewise.

Activity 2: FACILITATE DISCUSSION OF PARTICIPANTS' EXPECTATIONS OF THE COURSE AND THE GOALS OF THE COURSE.

Methodology: Group discussion
Lecturette
Handout 1A

Ask participants what they hope to gain from this physical fitness course. Refer participants to Handout 1A.

Activity 3: FAMILIARIZE PARTICIPANTS WITH AN OVERVIEW OF THE COURSE.

Methodology: Lecturette
Handout 1B

Refer participants to Handout 1B. Explain the content and format of the classes as described in the handout.

Activity 4: DISTINGUISH BETWEEN THE TWO MAIN TYPES OF EXERCISE AND FACILITATE DISCUSSION OF THE BENEFITS OF REGULAR EXERCISE, WITH EMPHASIS ON THE ELDERLY.

Methodology: Group discussion
Handout 1C

1. Stretching and strengthening exercises designed to increase flexibility and strength and improve muscle tone.
2. Aerobic exercises to strengthen the heart and cardiovascular system. To illustrate the distinction, ask the group for examples of each type of exercise.

Ask participants what they perceive to be the benefits of regular exercise, particularly for people of their age. Encourage them to include psychological and social benefits as well as physical benefits.

After they have responded, explain the benefits in Handout 1C that they did not mention.

Finally, refer to Handout 1C.

Activity 5: INTRODUCE AND PRACTICE STRETCH BREAK MOVEMENTS.

Methodology: Exercise session
Handout 1D

These exercises are a simple stretching routine. They may be used as a stretch break or as a warmup for further stretching exercises or aerobic exercise.

The facilitator's enthusiasm and sense of pleasure in exercise are vital at this point. If possible, participants should stand. Those not able to stand can participate from a seated position. Lead the group through all exercises listed in Handout 1D.

After the exercises, note the refreshed atmosphere and how the routine served as an effective stretch break. Refer participants to Handout 1D, suggesting that it be used as a guide for practicing the routine at home during the following week.

Activity 6: DESCRIBE FIVE NORMAL RESPONSES TO EXERCISE AND SEVEN RESPONSES INDICATING INAPPROPRIATE EXERCISE.

Methodology: Lecturette
Handout 1E

Refer participants to Handout 1E. Summarize its contents to the group.

Activity 7: DISCUSS SITUATIONS IN WHICH CAUTION IN EXERCISE IS ADVISABLE, AND PASS OUT THE "RELEASE OF LIABILITY STATEMENT" AND THE "PERSONAL RELEASE OF LIABILITY STATEMENT" (OPTIONAL).

Methodology: Lecturette
Release of Liability Statement
Personal Release of Liability Statement

Explain that anyone with heart problems, high blood pressure, chronic obstructive pulmonary disease, diabetes, or arthritis, or anyone thirty percent overweight should see a health care specialist before beginning an exercise program.

Two medically related liability release forms are used in the course. One is for participants who have been required by staff to secure a medical release in order to participate or who have doubts about their ability to engage in active exercise; this is the "Release of Liability Statement." The second form, the "Personal Release of Liability Statement," is for all other participants.

Distribute the appropriate form to each individual. Request that it be signed and returned at the next session.

Note that any health care specialist desiring more information about the program before signing the statement are welcome to call for information. Repeat the phone number.

Activity 8: EXPLAIN THE DAILY PHYSICAL ACTIVITY LOG AND ENCOURAGE ITS USE.

Methodology: Lecturette
Handout 1F

Refer participants to Handout 1F. Explain that it is a one-week physical activity survey designed to increase awareness of how much physical activity an individual is presently engaged in, and his/her feelings about the level of activity. Based on this information, a plan to increase activity at a safe rate can be developed. The survey may be repeated to assess programs and re-evaluate goals.

Suggest that the participants look over the form and ask any questions they may have as they begin to use it. Provide a hypothetical example to test their understanding of the use of the log.

Activity 9: EVALUATE THE SESSION.

Methodology: Group discussion
Flip chart

The purpose of the evaluation is to improve the course by incorporating participants' feedback. To elicit evaluatory comments, ask the group, "What were the most useful aspects of this session?" and "How could the session have been improved?" It is important that the facilitator be receptive to constructive criticism.

Activity 10: PREVIEW THE AGENDA FOR SESSION 2.

Methodology: Flip chart

The agenda for Session 2 should already be written on the flip chart. Briefly read it aloud.

RELEASE OF LIABILITY STATEMENT

TO ______________________________ .
health care specialist's name

I, ______________________________, am interested in participating in
your name

the ______________________________ exercise program, which includes stretching and aerobic exercises.

FOR THE MEDICAL PROVIDER

Please check one:

_______ Insofar as I am aware, there are no physical or mental contraindications to his/her full participation.

_______ The following are specific contraindications to his/her full participation. (please list):

____

____

____

____

____

____

Medical Provider _______________________
Date _______________________

Please send to:

If you need further clarification of our exercise program, please call _______________________ at _______________________, on _____________.

PERSONAL RELEASE OF LIABILITY STATEMENT

I, ___________________________, am participating in the exercise program without consent of my health care specialist. The activities, including aerobics and stretching exercises, have been explained to me and I fully understand what is involved in participation. I release the _________ _________ project from liability and accept that I am participating at my own risk.

Signed _______________________
Date _______________________

Witness_______________________
Date _______________________

HANDOUT 1A

THE GOALS OF THE CLASS

We would like each of you to experience the benefits of regular exercise. These benefits may include achieving greater flexibility and strength, improved endurance, improved circulation, increased energy and vitality, more restful sleep, more effective digestion of food, a lessening or elimination of depression or nervous tension, and an improved appearance and more positive self-image.

We also want you to have fun and experience pleasure in the physical activities you choose. We will help you obtain the kind and amount of physical activity that often results in improved health. Each of you will have the opportunity to slowly and gently increase your level of physical activity. How much you exercise will be determined by your present fitness level and your personal goals.

What We'd Like to Help You Accomplish and How

1. Increase flexibility and strength of muscles, joints, and bones by learning and practicing various stretching exercises.
2. Increase cardiopulmonary fitness by regularly engaging in aerobic exercises such as brisk walking, dancing, swimming, jogging, water exercise, and biking.
3. Gain a basic understanding of the physical effects of exercise through presentations, discussions, and readings.
4. Follow your progress by keeping records.
5. Strengthen your commitment to make exercise a regular part of your daily life by participating in group discussions of personal progress.
6. Assist in the development of the exercise program by sharing your ideas, selecting activities, and leading portions of class sessions when appropriate.

HANDOUT 1B

PHYSICAL FITNESS COURSE OVERVIEW

We will begin the classes by discussing what each participant would like to gain during the physical fitness course. We will help you develop a realistic plan for improving your fitness.

We will practice stretching exercises and provide information about the physical benefits of the movements. We will teach you how to increase the strength of your heart, lungs, and circulatory system through aerobic exercise. Guidelines, recommendations, and precautions for aerobic exercising will be discussed. Each person will be encouraged to exercise daily.

Most classes include presentation of new material, participation in stretching exercises and discussion about personal progress. We hope to create an atmosphere that is fun, noncompetitive, and supportive of change.

HANDOUT 1C

BENEFITS OF REGULAR EXERCISE

People who exercise regularly find that exercise:

—energizes
—helps you cope with stress and tension
—helps you to relax
—tones your muscles
—controls your appetite
—burns off calories
—improves self-image
—helps combat insomnia

Exercise is good for every body!!

HANDOUT 1D

STRETCH BREAK MOVEMENTS

Move every part of the body, every day, in every way. Tune in to your body. Be gentle with stretching. Do only as much as is comfortable for you. Be responsible for the health and safety of your own body.

Daily Routine for Stretching and Moving

1. Pucker up face and mouth, squint eyes. Yawn and relax face muscles.

2. Do head rolls, rotating head gently, first in one direction and then the other.
3. Move shoulders in a circular motion, forward and up, back and down.
4. Rotate hands at wrists in both directions. Move arms forward and backward at shoulder level. Swing arms in full circles to the front and then out to the sides.
5. Twist the trunk side to side allowing the head (and eyes) and arms to follow naturally.
6. Move legs forward and backward while holding on to a table or chair.
7. Flex each ankle—point heel, then toe. Rotate feet in each direction.
8. Gently bounce body all over, relaxing your entire body.
9. Take a moment to feel the energy flowing in your body. Proceed with any other exercises you want to do.

Adapted from an exercise routine developed by Dr. John McCamy. *Human Life Styling.* NY: Pocket Books, 1975.

HANDOUT 1E

RESPONSES TO EXERCISE

Some feelings and sensations occur as a normal reaction to exercise. These responses indicate your body is adapting to exercise and getting into shape. Normal responses include:

—Increased depth and rate of breathing
—Increased heart rate
—Feeling or hearing your heart beat
—Mild to moderate sweating
—Mild muscle aches and tenderness during the first weeks of exercise

Other sensations are not a normal reaction to exercise and may indicate you are overexerting, not exercising correctly, or that you have physical limitations that need to be discussed with a health care specialist. These responses include:

—Severe shortness of breath
—Wheezing, coughing, or other difficulty in breathing

—Chest pain, pressure, or tightness
—Lightheadedness, dizziness, fainting
—Cramps or severe pain or muscle aches
—Severe, prolonged fatigue or exhaustion after exercise
—Nausea

Learn to "tune in" to your body. As you pay attention to how your body feels, you will soon become your own best guide as to what you can do safely. If you have been relatively inactive lately, proceed *slowly* and *gradually*.

HANDOUT 1F

DAILY PHYSICAL ACTIVITY LOG

Name (please print) ____________________
Date log started ____________________

For each day in the following week, please record the type, duration, and frequency of all your physical activities. There are many different types of physical activities, and it is important that you list all types. For example, some activities are quite strenuous and increase your heart rate (e.g., running, swimming, racket sports, dancing). Others are not as strenuous but increase muscle flexibility and body control (e.g., yoga, stretching). A third type of physical activity increases strength and stamina (e.g., weight lifting and isometric exercise). Other examples of activities are bicycling, stationary bicycling, water exercise, swimming, jumping rope, gardening, walking, and jogging. You may engage in other physical activities, too, which you should list.

Under the column labeled "type of activity," list all physical activities engaged in each day of the next week. Under the column labeled "total time spent," record the total time spent that day doing the activity. Under the column labeled "number of times," indicate the number of times you engaged in the activity during that day. Some of you might walk 20 minutes in the morning and 30 minutes in the afternoon. If this applies to you, you would list walking under "type of activity," 50 minutes under "total time spent," and 2 under "number of times."

DAILY PHYSICAL ACTIVITY LOG

Name (please print) ______________________

	Type of Activity	Total Time Spent	Number of Times	Attitudes or Feelings
Monday				
Tuesday				
Wednesday				
Thursday				
Friday				
Saturday				
Sunday				

SESSION 2: BASIC STRATEGIES FOR DEVELOPING A PHYSICAL FITNESS PROGRAM

Session Agenda

1. Preview the agenda for the session, make announcements, and collect the "Release of Liability Statements" and "Personal Release of Liability Statements" distributed in Session 1.
2. Lead the group in reviewing the content of Session 1.
3. Facilitate discussion of barriers to exercising, and of ways to overcome these barriers as well as positive cues that encourage exercise.
4. Introduce basic stretching exercises to be done while standing, and lead the group in practicing them.
5. Provide nine practical tips for exercising.
6. Introduce the Affirmation of Health form, explain its purpose, and guide participants in filling it out for the following week's goal-related activity.
7. Evaluate the session.
8. Preview the agenda for Session 3.

Handouts

Handout 2A - Exercises Done While Standing
Handout 2B - Practical Tips for Exercising
Handout 2C - Affirmation of Health
Handout 2D - Sample Affirmation of Health

Materials Needed

Flip chart
Magic marker
Session agenda printed on flip chart
Agenda for Session 3 printed on flip chart
Daily Physical Activity Log

Activity 1: PREVIEW THE AGENDA FOR THE SESSION, MAKE ANNOUNCEMENTS, AND COLLECT THE "RELEASE OF LIABILITY STATEMENTS" AND "PERSONAL RELEASE OF LIABILITY STATEMENTS" DISTRIBUTED IN SESSION 1.

Methodology: Flip chart

Activity 2: LEAD THE GROUP IN REVIEWING THE CONTENT OF SESSION 1.

Methodology: Group discussion

Highlight the two main types of exercise and the situations in which caution in exercise is advisable.

Ask who practiced the stretch break routine during the week, how frequently, and with what effects. Strongly praise any affirmative responses.

Activity 3: FACILITATE DISCUSSION OF BARRIERS TO EXERCISING, AND OF WAYS TO OVERCOME THESE BARRIERS.

Methodology: Group discussion
Daily Physical Activity Logs
Small group discussion

Refer to the Daily Physical Activity Logs which participants kept during the past week. Ask who was surprised to realize how much or how little exercise he/she engaged in. Ask who was satisfied with his/her activity level, and inquire who would like to be more physically active.

Instruct the group to divide into small groups of four, and to spend a few minutes identifying reasons why they do not exercise as much as they would prefer—in other words, to identify barriers to exercising more. Suggest that the attitudes recorded in their logs might provide insight.

In addition to discussing barriers to exercise, ask successful exercisers to share how they manage their time and environment so as to encourage exercise.

Reconvene the group and discuss their findings. Ask each person to share a barrier. Encourage the group as a whole to suggest practical ways of overcoming each barrier. Time may prohibit an in-depth discussion of each barrier; focus on those that appear most common.

Activity 4: INTRODUCE BASIC STRETCHING AND STRENGTHENING EXERCISES TO BE DONE WHILE STANDING, AND LEAD THE GROUP IN PRACTICING THEM.

Methodology: Exercise session
Handout 2A

The fourth class will focus on more stretching and strengthening exercises. The intent of today's session is to introduce some basic exercises that might be used both by participants at home in the coming week and as warm-up exercises in Sessions 3 and 4.

Lead them through all the exercises on Handout 2A. Again, the enthusiasm and genuine enjoyment of exercise expressed by the facilitator is vital.

After completing the routine, distribute Handout 2A for reference during the week.

Activity 5: PROVIDE NINE PRACTICAL TIPS FOR EXERCISING.

Methodology: Reading aloud
Handout 2B

Refer participants to Handout 2B. Read through each tip aloud, or ask a different participant to read each one.

Activity 6: INTRODUCE THE AFFIRMATION OF HEALTH FORM, EXPLAIN ITS PURPOSE, AND GUIDE THE PARTICIPANTS IN FILLING IT OUT FOR THE FOLLOWING WEEK'S GOAL-RELATED ACTIVITY.

Methodology: Lecturette
Small groups
Handout 2C
Handout 2D

Refer participants to Handout 2C. Explain that each person will use the form to make weekly agreements with him/herself to do some physical activity. Each weekly activity or short-term goal will contribute to the long-term goal of beginning or improving, and maintaining, an exercise program that will contribute to a healthier lifestyle.

Note that the bottom portion of the page provides an opportunity for enlisting the support of at least one other person, as well as for rewarding oneself for having completed the activity.

Refer to Handout 2D. Explain that it is an example of a completed form. Ask participants to look it over and to ask questions.

Direct participants to move into groups of four. With the help of the other small group members, each person should decide upon a physical activity for the following week, and using the bottom portion of the form, devise a plan for carrying it out. The activity may be either of the stretching exercise routines learned thus far in the course. Explain that everyone will have a chance, in the following week, to report back to the group concerning the week's activity.

Activity 7: EVALUATE THE SESSION.

Methodology: Group discussion
Flip chart

Same as Session 1, Activity 10.

Activity 8: PREVIEW THE AGENDA FOR SESSION 3.

Methodology: Flip chart

The agenda for Session 3 should already be written in the flip chart. Briefly read it aloud.

HANDOUT 2A

EXERCISES DONE WHILE STANDING

Keep knees slightly flexed and feet shoulder-width apart for these exercises.

1. *Yawn & Stretch:* stretch arms to the sky, rise on toes, move arms out to the side and down.
2. *Picking Grapes:* inhale while raising arms to the sky alternately four times, exhale, and lower arms, head, and upper torso.

3. *Side Stretch:* with arms out to the side, raise one arm over your head and bend to the other side. Reverse sides.
4. *Shoulder Rolls:* inhale while bringing shoulders forward and up. Exhale while bringing shoulders back and down.
5. *Arm Circles:* raise arms forward and above shoulder height. Inhale as arms circle four times. Exhale as the arms circle out.
6. *Hand Circles:* raise forearms to shoulder height. Inhale and rotate wrists inward four times. Exhale and rotate wrists outward in a circle.
7. *Hand Squeezes:* make fists with hands and release.
8. *Windmill:* inhale and move arms as though you are doing the backstroke in water. Exhale and reverse arm movement.
9. *Scythe:* gently swing torso side to side, turning the head and looking behind. Arms should swing freely.
10. *Hip Rotations:* hands on hips. Tilt the hips forward, and then back. Repeat. Tilt the hips to the right, then left. Repeat.
11. *Hip Circles:* hands on hips. Move the hips in a smooth circle in one direction. Reverse direction.
12. *Turnstile:* lift the leg forward with knee bent, rotate outward and inward. Lower leg. Repeat with other leg.
13. *Neck Series* (with shoulders relaxed):
    - A) *Head Roll:* gently roll your head in a circle in each direction. Avoid jutting your chin out as your head rolls to the back.
    - B) *Side to Side:* turn your head to one side, then the other, in a "no" movement.
    - C) *Camel:* inhale moving the head forward and down. Exhale bringing the head back and up.
14. *Ankle Rotations:* lift one leg slightly in front of you. Rotate the foot in a circle to the right and then to the left. Alternate feet.
15. *Foot Flexion & Extension:* raise one leg in front of you. Alternately point the toes and flex the foot. Alternate feet.
16. *Toe Writing:* raise one leg in front of you. Write your first name in the air with your foot. Alternate and write your last name with your other foot.
17. *Posture:* align your spine beginning with your feet together, tucking your pelvis under, flattening your abdomen, and opening your chest with shoulders back and down and head tall.
18. *Mountain:* stand with spine straight, feet slightly apart. Inhale deeply and slowly. Lift arms over your head. Exhale and lower arms.

HANDOUT 2B

PRACTICAL TIPS FOR EXERCISING

1. Be sure you have loose and comfortable clothing and proper equipment for the activity you choose. (For example, comfortable shoes for walking or goggles and ear plugs for swimming.
2. Do not exercise on a full stomach. Wait at least two hours after a meal.
3. If possible, find a regular time of the day to exercise or a consistent schedule to follow.
4. Exercise with a friend or family member, or join a class or club.
5. Pick an exercise you enjoy.
6. Proceed slowly and gradually. Chart your progress. Tell your friends how well you are doing!
7. If you exercise outdoors, pick a scenic, relatively unpolluted place to exercise.
8. Read articles or books about the activity you do, written by people who are enthusiastic. Their excitement may rub off onto you!
9. Use positive imagery:

   —See yourself as vital and healthy
   —See your arteries free of fatty deposits
   —See yourself losing weight
   —See yourself energized, walking in the fresh air
   —See yourself warm and relaxed

Adapted from: *The Family Guide to Exercise and Nutrition,* prepared by The Stanford Heart Disease Prevention Program.

HANDOUT 2C

AFFIRMATION OF HEALTH

Name ______________________

1. The major goal I want to accomplish in the next ____ weeks is:

2. How I plan to accomplish this goal is:

Week	Activity	Where	When	Times Per Week	With Whom

3. In order to help me follow through with my activity, I will:

   ______ Keep a journal of my thoughts, feelings, and reactions.

   ______ Keep records on a Daily Activities Log.

   ______ Invite a friend or family member to work with me in my effort to change: the way this person will help me is:

   ____

   ____

   Invite someone from the class to work with me in my effort to change: the way this person will help me is:

   ____

   ____

   ______ Reward myself by ____

   ____

HANDOUT 2D

AFFIRMATION OF HEALTH

Name Katharine Maguire

1. The major goal I want to accomplish in the next 4 weeks is:
   To walk 3 miles, 4 times per week, in less than one hour.

2. How I plan to accomplish this goal is:

Week	Activity	Where	When	Times Per Week	With Whom
1	walking	around block	morning	4	my neighbor
2	walking	park & back	afternoon	4	classmate
3	walking	½ way around lake	afternoon	4	classmate
4	walking	all around lake	afternoon	4	classmate

3. In order to help me follow through with my activity, I will:

✓ Keep a journal of my thoughts, feelings, and reactions.

______ Keep records on a Daily Activities Log.

✓ Invite a friend or family member to work with me in my effort to change: the way this person will help me is: by walking with me 3 times per week.

✓ Invite someone from the class to work with me in my effort to change: the way this person will help me is: by walking around the lake with me once a week.

✓ Reward myself by reading my favorite novel after I walk.

SESSION 3: AEROBIC EXERCISE

Session Agenda

1. Preview the agenda for the session and make announcements.
2. Facilitate sharing of the Affirmation of Health.
3. Define aerobic exercise.
4. Review the physical, emotional, and social benefits of regular exercise as identified in Session 1, Activity 4.
5. Teach participants how to begin an aerobic exercise program.
6. Teach participants how to measure pulse rates and to calcu-

late target heart rates; clarify the function of heart rate measurements in an aerobic exercise program.

7. Emphasize the importance of warming up and cooling down as part of aerobic exercise.
8. Discuss walking as one of the best aerobic exercises for older adults.
9. Develop personal aerobic exercise programs.
10. Evaluate the session.
11. Preview the agenda for Session 4.

Handouts

Handout 3A – What is Aerobics?
Handout 3B – How do I Begin an Aerobic Exercise Program?
Handout 3C – Measuring Your Pulse Rate
Handout 3D – Walking
Handout 1C – Benefits of Regular Exercise
Handout 2C – Practical Tips for Exercising

Materials Needed

Flip chart
Magic markers
Session agenda printed on flip chart
Agenda for Session 4 printed on flip chart
Affirmation of Health forms
A clock with second hand, visible to all participants

Activity 1: PREVIEW THE AGENDA FOR THE SESSION AND MAKE ANNOUNCEMENTS.

Methodology: Flip chart

Briefly describe each item. Make any necessary announcements and invite participants to do the same.

Activity 2: FACILITATE SHARING OF THE AFFIRMATION OF HEALTH FORM.

Methodology: Group discussion
Affirmation of Health forms

A detailed description of how to effectively facilitate sharing is provided in the Stress Management module, Session 3, Activity 3. That description also applies to sharing in this exercise module.

Activity 3: DEFINE AEROBIC EXERCISE.

Methodology: Group discussion
Handout 3A

Explain the meaning of "aerobic" and "aerobic exercise" as defined in the first two paragraphs of Handout 3A.

To illustrate these definitions, ask the group to suggest exercises that might be considered aerobic. Discuss each response, pointing out why it is or is not aerobic. Activities which should be identified as aerobic are jogging, running, swimming, bicycle riding, and brisk walking. If any of these are omitted by the group, the facilitator should identify them.

Finally, refer participants to Handout 3A.

Activity 4: REVIEW THE PHYSICAL, EMOTIONAL, AND SOCIAL BENEFITS OF REGULAR EXERCISE AS IDENTIFIED IN SESSION 1, ACTIVITY 4.

Methodology: Group discussion
Handout 1C

Refer participants to Handout 1C which discusses the benefits of exercising. Point out that a program of regular exercise is most beneficial if it includes both of the two main exercise types: aerobic exercise and stretching and strengthening exercises.

Activity 5: TEACH PARTICIPANTS HOW TO BEGIN AN AEROBIC EXERCISE PROGRAM.

Methodology: Lecturette
Handout 3B

Refer participants to Handout 3B. Look over the aerobic exercise checklist with the group. Encourage participants to express and discuss any questions they may have.

Activity 6: TEACH PARTICIPANTS HOW TO MEASURE PULSE RATES AND TO CALCULATE TARGET HEART RATES; CLARIFY THE FUNCTION OF HEART RATE MEASUREMENT IN AN AEROBIC EXERCISE PROGRAM.

Methodology: Lecturette or reading aloud
Handout 3C
Practice measuring pulse rates

Refer participants to Handout 3C. Explain its contents to the group, read it aloud, or ask a participant to read it aloud. Encourage questions. (Omit, at this point, the final paragraph on warming up and cooling down.)

Guide participants step-by-step in measuring their pulse rates. Carry out the technique a number of times with facilitator as timer and then with participants timing themselves. Emphasize the importance of doing the measurement immediately upon stopping an aerobic activity.

Activity 7: EMPHASIZE THE IMPORTANCE OF WARMING UP AND COOLING DOWN AS PART OF AN AEROBIC EXERCISE ROUTINE.

Methodology: Reading aloud
Handout 3C

Read aloud the final paragraph of Handout 3C, or request a participant to do so. The stretching routines learned in Sessions 1 and 2 are ways to warm up and cool down for aerobic exercise.

Activity 8: DISCUSS WALKING AS ONE OF THE BEST EXERCISES FOR OLDER ADULTS.

Methodology: Lecturette
Group discussion
Handout 3D

Ask how many participants walk regularly and how often.

Refer participants to Handout 3D and review its contents.

Activity 9: FACILITATE DEVELOPMENT OF PERSONAL AEROBIC EXERCISE PROGRAMS.

Methodology: Small groups
Lecturette
Affirmation of Health forms
Handout 2C

The group now has some of the knowledge needed to begin a good personal aerobic exercise program.

Direct participants to divide into small groups of 3 or 4; close friends or family members should be in the same small group. Direct the groups to use the Affirmation of Health to develop an individual aerobic exercise plan for each group member, following these guidelines:

—Item #1 on the Affirmation of Health (the major goal) should be to be begin and/or maintain a regular exercise program including both aerobic exercise and stretching/strengthening exercise.
—The activity for this week will be recorded as ''week 2'' (Session 2's activity was recorded as ''week 1'') and should be some form of aerobic activity.
—The activity chosen should be one which is most congruent with the individual's preferences, abilities, and resources (i.e., a person without access to a swimming pool should not choose swimming). ''Times per Week'' and ''With Whom'' should also be feasible.
—Each person should choose those tools in item #3 that he/she considers most helpful. Encourage everyone to invite either a friend, family member, or classmate to work with them in accomplishing the activity. This person may be someone from the small group. Strongly encourage everyone to plan a healthful self-reward.
—Anyone choosing to keep a daily activities record may use the log distributed in Session 1.

Suggest that everyone refer to Handout 2C, ''Practical Tips for Exercising,'' before developing the individualized plan. The plans should be developed and recorded on the Affirmation of Health before the end of this session. Remind the group that in Session 4, they will have the opportunity to share with the group their goal-related activity during the week.

Activity 10: EVALUATE THE SESSION.

Methodology: Group discussion
Flip chart

Same as in previous sessions.

Activity 11: PREVIEW THE AGENDA FOR SESSION 4.

Methodology: Flip chart

Same as in previous sessions.

HANDOUT 3A

WHAT IS AEROBICS?

The word "aerobic" technically means "with air." An aerobic exercise is one which increases the heart and breathing rates for a sustained period. It requires continuous exertion rather than frequent stops and starts.

To be effective in improving heart and lung fitness, the activity must raise the heart rate to a certain level (see the chart in Handout 3C) and maintain that level for 15-20 minutes. This activity must be engaged in 3-4 times per week, preferably every other day. If a person goes 3-4 days without exercise, some of the benefits are lost.

Exercises suitable to improve heart and lung fitness are brisk walking, jogging, running, swimming, and bicycle riding. Dancing is suitable if the activity is vigorous and sustained. Activities which are *not* aerobic include weight lifting, yoga and stretching exercises, calisthenics, tennis, bowling, and gardening.

HANDOUT 3B

HOW DO I BEGIN AN AEROBIC EXERCISE PROGRAM?

First, review this checklist. If you answer "yes" to any of these questions, consult your health care specialist before beginning an aerobics program.

1. Has a doctor ever told you that you have heart trouble?
2. Do you have any physical condition which limits your movement and could be aggravated by exercise?
3. Are you more than 30 lbs. overweight?
4. Do you get short of breath climbing one flight of stairs?
5. Do you ever have chest or heart pains, especially when you are physically active?
6. Do you have high blood pressure or hypertension?
7. Do you often feel faint or have dizzy spells?
8. Are you over 40 and not used to vigorous exercise?

If you answered "no" to all these questions, you are probably able to safely begin a regular exercise program. If you have any doubt about your ability to exercise regularly, check with your health care specialist.

It is important to begin slowly and easily with an activity you think you will enjoy. If you have not been exercising regularly, it will take time for your heart and muscles to adjust to the new demands. Proceed with patience—if you expect too much of yourself too quickly, you may end up sore and vowing never to exercise again! If you have a cold or any other illness, wait until you are better to begin exercising.

Remember—it may take you a number of weeks of gradually increasing your activity before you can exercise 15 - 20 minutes with your heart rate in the recommended aerobic range. Charting your progress is a way to recognize that you are indeed moving toward your goal.

How Much Exercise Is Enough?

If your exercise routine is adequate to improve your heart and lung fitness, you will experience a number of physical changes while you work out. Your heart and breathing rates will be noticeably increased and you may feel or hear your heartbeat. You will likely be sweating. Your muscles will be tired, though not exhausted, after your workout. Your muscles may be somewhat sore for the first few weeks after you begin exercising.

Adapted from: *The Family Guide to Exercise and Nutrition,* prepared by The Stanford Heart Disease Prevention Program.

HANDOUT 3C

MEASURING YOUR HEART RATE

As you exercise, your heart rate increases to meet the greater demand for blood supply to your muscles. To increase your aerobic fitness, your heart rate should be raised to within the range listed for your age on the following chart. This is a heart rate to aim for during vigorous activity.

Heart Rates for Exercising

Age	Heart Rate Range (Beats/Minute)	Maximum Heart Rate (Beats/Minute)
15	148–180	205
20	144–176	200
25	140–171	195
30	137–167	190
35	133–163	185
40	130–158	180
45	126–154	175
50	122–150	170
55	119–145	165
60	115–141	160
65	112–136	155
70	108–132	150
75	104–127	145
80	100–122	140
85	96–118	135
90	91–113	130

Adapted from *The Family Guide to Exercise and Nutrition*, prepared by The Stanford Heart Disease Prevention Program.

The heart rate range for exercising is fairly wide because individual heart rates differ to some extent. The listed range is about 60-80% of your aerobic capacity, which is adequate to improve fitness. Exceeding the range is not recommended—this unnecessarily increases risk of injury.

Also, note the maximum heart rates listed. These are "average" values for each age and individual variation may be great. These figures indicate that as we grow older, the highest heart rate which can be reached during a vigorous exercise effort falls. Thus, an older person may expend the same degree of effort in exercising as a younger person but her or his heart rate may be considerably less. That person is at the same percentage of her or his maximum capacity as the younger person.

How Much Exercise Is Necessary to Gain Some Benefit?

One study indicates that men (unfortunately, the study did not test women) of average physical fitness, in their 60's and 70's, could improve in fitness by raising their heart rates above 98 and 95 respectively. Even well-conditioned men in these age brackets needed to exceed 106 and 103, respectively, to gain some benefit. This was the heart rate immediately after exercise. The study concluded that, "for all but the highly conditioned older men, vigorous walking, which raises heart rate 100 to 120 beats per minute for 30 – 60 minutes daily will bring some, though not perhaps the best, improvement in heart and lung function."*

Taking Your Pulse

You can measure your heart rate in beats per minute by counting your wrist pulse for 6 seconds and adding a zero to that number. To be accurate, start counting with zero as the six-second period begins and stop counting at the end of six seconds. If you count 11 beats in the six-second period, your heart rate in beats per minute is 110. To find your wrist pulse, hold your arm with your palm up, facing you. Bend your hand slightly away from you. Place the tips of your index and middle fingers of your other hand on the center of your wrist, slightly above the bend at your wrist. Slide your fingertips toward the outside of your wrist (the thumb side) and down over two tendons and into a soft depression between the tendons and the bone that runs along the thumb side of your wrist. Apply gentle pressure in that depression. Be sensitive to the pulsations there; it may take

---

*De Vries, H.A. *Geriatrics*, 1971, *26*, 94.

a little time to learn to feel your wrist pulse. Sometimes it is easier to find your pulse on one wrist than the other. Experiment!

In order to get an accurate exercise pulse (your pulse during exercise), you must take your pulse after you have been exercising for at least five minutes at the rate you wish to maintain during your workout. In other words, if you swim for exercise, you should be swimming at your workout pace for at least five minutes before taking your pulse. This allows your heart to reach the maximum rate for that level of activity. Since the heart rate slows quickly when you stop exercising, it is important to take your pulse immediately after stopping the activity.

Helpful Hints

You will need to have a watch with a second hand. It may help to have a friend who can time you while you count your pulse. Learning to take an accurate exercise pulse does take practice, so do not be alarmed if the pulse reading varies considerably as you learn the proper technique.

Warm Up and Cool Down

For each exercise session, it is important to warm up your body by stretching and/or beginning your activity at a very slow pace and increasing gradually. This will help you avoid injury and increase your enjoyment of the activity.

It is equally important to cool your body down after a vigorous workout to ease the transition to resting. During the last five minutes of your workout, gradually slow your rate of movement, allowing your breathing and heart rates to return to normal. Ending a workout with stationary stretching exercises will reduce the chance of being stiff or sore the following day.

HANDOUT 3D

WALKING

Walking is one of the best and safest all-around exercises for older adults. Brisk walking is an aerobic activity that strengthens the heart and lungs and improves endurance. No special skill is involved, there is no charge, it can be done almost anywhere, in solitude or with companions, and it has a lower injury rate than most other exercises.

In order to receive conditioning benefits from walking you must walk at a pace that deepens your breathing and increases your heart rate. Use your aerobic heart rate range as a guide to determine how briskly you must walk for best results. (Increasing your pace from 3 to 5 mph will increase the calories you burn from 66 up to 124 calories per mile!)

A good walking workout depends upon stepping up your pace, increasing your distance, and walking more often. Here are some tips to help you get the most out of walking:

1. Walk steadily and briskly.
2. Breathe deeply through your nose or mouth, whichever is more comfortable.
3. Check your pulse (as suggested in the aerobics section) to monitor your heart rate.
4. Wear comfortable clothes that allow you to take long, easy strides. In cold weather wear several layers of light clothing so you can remove layers as you warm up. A scarf and cap is crucial in very cold weather.
5. Lean forward slightly when walking up hills, and be sure to breathe deeply.
6. Land on your heel and roll forward to step off the ball of your foot. Soreness may result from walking flatfooted or only on the balls of your feet.
7. Keep your head erect, shoulders back and relaxed, and your back straight. Your toes should point straight ahead and your arms should swing freely.
8. Wear comfortable shoes which provide good support and don't cause blisters or calluses. Your shoes should have uppers made of materials which "breathe" such as leather or nylon mesh and the soles should be made of non-slip material.

Adapted from: *Pep Up Your Life: A Fitness Book for Seniors.* Printed by The Travelers Insurance Companies, Hartford, Connecticut 06115.

SESSION 4: STRETCHING AND STRENGTHENING EXERCISES

Session Agenda

1. Preview the agenda for the session and make announcements.
2. Facilitate sharing of the Affirmation of Health forms.
3. List the benefits of stretching and strengthening exercises and discuss general precautions.
4. Lead the group in doing exercises lying down, seated, using a support, and on hands and knees.
5. Discuss and demonstrate exercises and other techniques for relieving low back problems.
6. Guide participants in incorporating stretching and strengthening exercises into the personal exercise programs begun in Session 3.

*7. Evaluate the Physical Fitness course.

Handouts

Handout 4A – Stretching and Strengthening Exercises
Handout 4B – How to Do Stretching and Strengthening Exercises
Handout 4C – Back Care

Materials Needed

Flip chart
Magic markers
Session agenda printed on flip chart
Exercise mats
Written course evaluation forms (found in Tips for Facilitators section)

Activity 1: PREVIEW THE AGENDA FOR THE SESSION AND MAKE ANNOUNCEMENTS.

Methodology: Flip chart

Same as in previous sessions.

---

*If this is *not* the final session of the course, Activity 7 should be altered to read "Evaluate the Session," and Activity 8 "Preview the Agenda for Session 5," should be added.

Activity 2: FACILITATE SHARING OF THE AFFIRMATION OF HEALTH FORMS.

Methodology: Group discussion
Affirmation of Health forms

Same as Session 3, Activity 2.

Activity 3: LIST THE BENEFITS OF STRETCHING AND STRENGTHENING EXERCISES AND DISCUSS GENERAL PRECAUTIONS.

Methodology: Group Discussion
Handout 4A

Refer participants to Handout 4A. Either read it aloud or request a participant to do so. Encourage participants to ask questions and to express reservations they may have.

Activity 4: LEAD THE GROUP IN DOING EXERCISES LYING DOWN, SEATED, USING A SUPPORT, AND ON HANDS AND KNEES.

Methodology: Exercise session
Handout 4B

The group has learned and practiced two basic stretching and strengthening exercise routines; Session 1 contained "Stretch Break Movements," and in Session 2 they learned 18 exercises to be done while standing.

This session will include stretching and strengthening exercises to be done lying down, seated, using a support, and on hands and knees.

If some members of the group are confined to wheel chairs, many of these exercises will be appropriate for them, including some of the lying down exercises, hands and knees exercises, as well as those to be done while seated. (In leading the exercises, be sure to specify those which are appropriate to persons confined to wheel chairs and encourage their participation.)

Provide exercise mats for all other participants.

Lead the group through the exercises in Handout 4B, demonstrating each and then asking participants to do likewise. As in previous sessions, sincere enthusiasm and evident enjoyment in the facilitator are important. Recorded music is also very helpful.

After completing all the exercises, encourage participants to notice how they feel and to comment on their feelings, both physical and emotional.

Refer participants to Handout 4B for use in practicing the exercises at home.

Activity 5: DISCUSS AND DEMONSTRATE EXERCISES AND OTHER TECHNIQUES FOR RELIEVING LOWER BACK PROBLEMS.

Methodology: Demonstration
Reading aloud
Group discussion
Handout C

Refer participants to Handout 4C. Demonstrate the five exercises designed to strengthen a back that has been weakened by a strain, defect, disease, or lack of exercise. Emphasize the importance of starting slowly, of not overdoing it, and of contacting a doctor if more than mild pain should occur.

Ask participants to refer to the second and third pages of the handout. Read through each of the helpful hints for reducing back strain.

Finally, read through the various home treatments (using heat and cold) for the relief of back ache listed on the third page of Handout 4C.

Encourage discussion of the exercises, the helpful hints and the home treatments.

Activity 6: GUIDE PARTICIPANTS IN INCORPORATING STRETCHING AND STRENGTHENING EXERCISES INTO THE PERSONAL EXERCISE PROGRAMS BEGUN IN SESSION 3.

Methodology: Lecturette
Affirmation of Health forms
Handout 2C
(Optional: small groups)

Ask participants to refer to their Affirmation of Health forms. A good exercise program includes both of the two main types of exercise (aerobic and stretching and strengthening).

Explain that since last week's activity began the aerobic exercise component, this week's activity will develop the stretching and strengthening exercise component. Again, emphasize the importance of choosing activities (short-term goals) that are attainable. Strongly encourage each person's plan to include both the help of a friend, family member, or classmate and a self-reward. Suggest reviewing the "Practical Tips for Exercising" (Handout 2B).

Allow the group to choose whether or not to break into small groups for filling out the Affirmation of Health for this week.

Activity 7: EVALUATE THE EXERCISE COURSE.

Methodology: Group Discussion
Written course evaluation forms (available in the "Tips for Facilitators" section)

Explain that the course evaluation by participants is an important means of improving it for future participants. Request that they be honest and open in their critiques.

Distribute the course evaluation form and ask that everyone fill it out and hand it in before leaving.

HANDOUT 4A

STRETCHING AND STRENGTHENING EXERCISES

Objectives

These exercises are designed to increase flexibility and strength, improve muscle tone, and enhance general body functioning.

General Precautions and Information

If you feel dizzy or faint, sick to your stomach, experience any chest tightness, pain, or severe shortness of breath while exercising, stop exercising immediately. If symptoms persist, see your health care specialist. If you experience muscle pain or cramping during any exercise, stop that exercise, relax the affected muscle, and gently rub it with your hands. Proceed more slowly with easier movements.

Always exercise with awareness of your body's limits. Remember to proceed gently as you warm up and allow a time to cool down at the end of each session. The number of times you do each exercise will depend on your individual fitness level and your personal fitness goals. Do not be discouraged if you can only do a few repetitions of an exercise. With regular practice you will be able to comfortably increase repetitions as your body strengthens and becomes more flexible.

Wear loose, comfortable clothing. Exercise either in bare feet or in non-slip socks, stockings, or slippers. Exercise in a well-ventilated room if possible. Always wait a few hours after eating to exercise.

Full, relaxed breathing is important to gain the most benefit from exercising. Some of the exercises suggest how to breathe. Follow those suggestions. Remember to take a relaxed, full breath between each exercise. If you feel slightly light-headed while breathing deeply, return to your natural breathing rhythm until the light-headedness has ended.

HANDOUT 4B

HOW TO DO STRETCHING AND STRENGTHENING EXERCISE

The following exercises are adapted from arica psychocalisthenics, McCamy's Movements and Yoga exercises. They include exercises you can do on the floor, on your hands and knees, and in a chair.

The exercises marked by an asterisk (*) strengthen the back and the abdominal muscles and help prevent low back problems. If you presently are experiencing low back problems, show these exercises to your health care specialist to be sure they are appropriate for your situation.

These exercises are a continuation of "exercises done while standing" (Handout 2A).

Exercises Done While Lying on Your Back on the Floor

*1. *Knee to Chest:* Raise one leg, bending the knee. Grasp leg below the knee with hands and gently pull your leg towards your chest. For further stretching, raise your head and bring your chin toward your knees.

2. *Angel Stretch:* Begin with legs together and arms to side. Inhale and open arms and legs out from the center of your body along the floor. Exhale and bring arms and legs back together.

*3. *Pelvic Tilt and Lift:* Place feet flat on floor close to buttocks, with knees in the air. Press small of back to the floor. Slowly raise pelvis off the floor, one vertebra at a time. Reverse the process to lower pelvis.

*4. *Abdominal Curl:* Position legs as above in exercise no. 1. Slowly lift head and shoulders toward knees with arms outstretched toward knees. Hold for a count of four. Lower your head and shoulders slowly. Relax and repeat. This is just like a sit-up but with your legs bent.

5. *Side to Side Rocking:* Bring knees to chest and clasp with your arms. Rock from side to side for a quick back massage.

*6. *Alternate Leg Lifts:* With legs out straight, inhale and lift *one* leg from the hip. Exhale and lower slowly.

Exercises Done While Lying on Stomach on the Floor

*1. *Cobra:* Place hands next to shoulders. Slowly and gently stretch head and shoulders forward and up. Hold for a few seconds and gently lower shoulders and head. Rest and repeat. Caution: if you have low back problems, check with your health care provider before doing this exercise.

2. *Gluteal:* Tighten buttocks to a count of four, bringing heels together. Hold for another count of four and release. Relax and repeat.

*3. *Leg Lifts:* Raise one leg at a time off the floor. Keep leg straight.

Exercises Done on Hands and Knees

*1. *Cat:* Slowly inhale and raise head, while letting belly drop slightly. Exhale and tuck head and buttocks under while raising back and rounding back towards the ceiling.

2. *Side Stretch:* Turn upper body to the right, looking around behind at feet. Return to center and repeat to the left.

3. *Lateral Arm Raise:* Slowly raise right arm to the side as far as is comfortable, following the movement of the hand with the head and eyes. Return to center and repeat with left arm.

Exercises Using a Chair, Wall or Other Support

1. *Pushout:* Stand an arm's width from the wall and place your hands against it, palms out. Slowly bend your arms and do a "pushup" against the wall.
2. *Squat:* Hold on to the back of the chair. Spread your feet shoulder-width apart. Bend your knees as you exhale. Inhale as you stand up straight.
3. *Rise on Toes:* In a smooth, slow motion, raise your body on your toes as you inhale. Lower body slowly as you exhale.
4. *Leg Swings:* Using a chair or wall for balance, swing your outside leg freely forward and back. Alternate your position and swing the other leg freely.
5. *Leg Raises:* Body as above. Using the outside leg, raise and lower it slowly to the front, side, and back. Shake the leg to relax it. Repeat with the other leg.

Exercises Done While Seated in a Chair

NOTE: In addition to these seven exercises, many of the previously listed exercises can be adapted to chair use. Exercises done while standing handout 2A (numbers 1-7 and 13-16) and exercises done on hands and knees (numbers 2 and 3).

1. *Lion:* Inhale deeply and, while exhaling with eyes and mouth wide open, stick tongue out and roar!
2. *Palm Stretch:* With arms outstretched, interlock fingers and turn hands away from the body. Stretch the fingers and palms. Bring arms over head and again stretch the fingers and palms. Place palms behind head. Reverse procedure—stretch palms overhead and then in front.
3. *Chest Expansion:* Place hands on shoulders, elbows in front of body. Touch elbows together in front of body. Open elbows wide and expand chest. Repeat.

*4. *Knee Raise:* Raise left knee and lower. Repeat with right knee.

5. *Knee Squeeze:* Lift left knee and clasp arms around knee and squeeze. Lower and repeat with right knee.
6. *Side Twist:* Turn torso to one side, head and eyes following. Return to center and repeat exercise, turning to the other side.

*7. *Backstretch:* Bend forward and then straighten up. Repeat, clasping your hands on your left knee and then straightening up. Repeat with the right knee.

Reminder: Take a few moments to breathe deeply and to consciously feel the effects of exercising your body. Remind yourself of the benefits of exercising and appreciate yourself for treating yourself so well!

HANDOUT 4C

BASIC EXERCISES FOR THE LOW BACK SYNDROME

Adapted from GROUP HEALTH COOPERATIVE OF PUGET SOUND materials.

These exercises are designed to strengthen a back that has been weakened by a strain, defect, disease, or a simple lack of exercise. Start the exercises slowly. Don't overdo it! Follow your doctor's instructions carefully. Consult your doctor if more than mild pain occurs. Begin the exercises as shown below.

Starting position for all exercises: knees and hips bent with back flat and neck comfortably supported; arms to the side, feet flat on the floor.

1.

Take in a deep breath, exhale slowly. Tighten the stomach and buttock muscles and hold the back flat against floor for a count of five. Relax. Repeat very slowly 5 times.

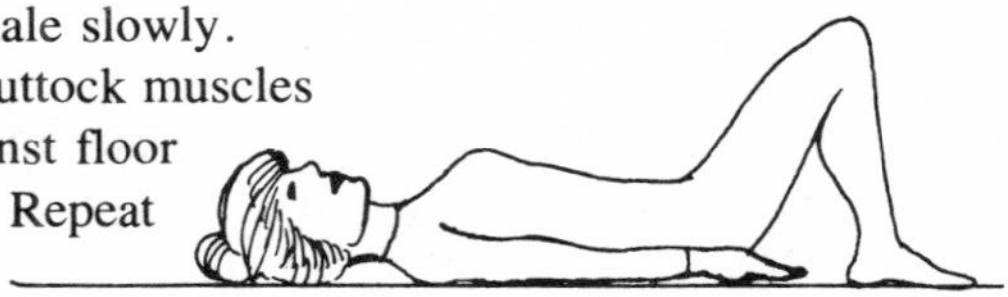

2.

With both hands on one knee, bring the knee up as near to the chest as possible. Return it slowly to the starting position. Relax. Repeat, alternating with each leg 10 times.

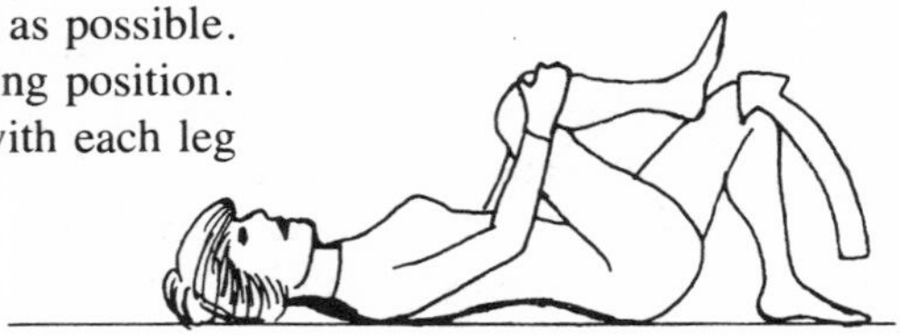

3.

Tighten the abdominal muscles and hold the back flat, then bring both knees up to the chest, grasp the knees with the hands and hold the knees against the chest about 30 seconds. Return to the starting position. Relax. Repeat 5 times.

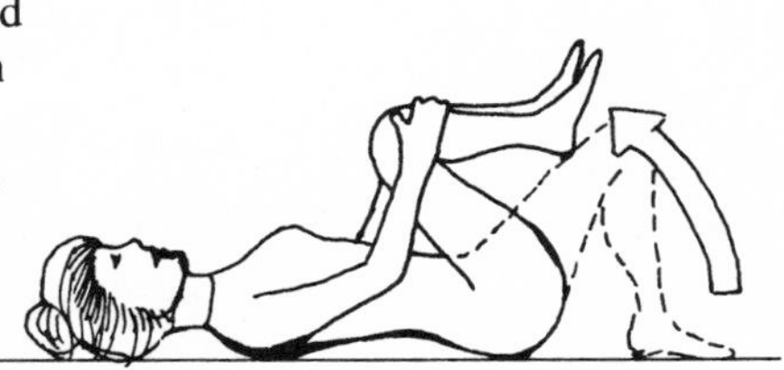

4.

Bring one knee to the chest. Straighten that knee, extending the leg as far as possible, bend knee and return to original position. Relax. Alternate with the opposite leg. Repeat 5 times. NOTE: This exercise is *not* recommended for patients with sciatic pain.

5.

Lie on back with knees bent, feet flat on floor. Pull up to a sitting position. Hold for a count of 5. Return to starting position. Relax. Repeat 5 times.

BACK CARE

Experience suggests that back pain is not cured by professional care alone; a painful back usually improves when it is well cared for by its owner 24 hours a day.

The most helpful thing we can do for your back is to show you how to take care of it yourself by learning new ways to stand, to sit, to lift, and even to sleep or rest. Take special note of the sleeping and resting positions below since it is possible to strain your back just by resting it improperly. A well positioned rest can be a very effective pain reliever.

At your doctor's request, the therapist may give you certain exercises to be done every day. The exercises are specifically designed to stretch and strengthen the muscles that support your back. You can combine daily exercise, good posture and sensible body mechanics in your own effective back care program.

RECOMMENDED RESTING OR SLEEPING POSITIONS

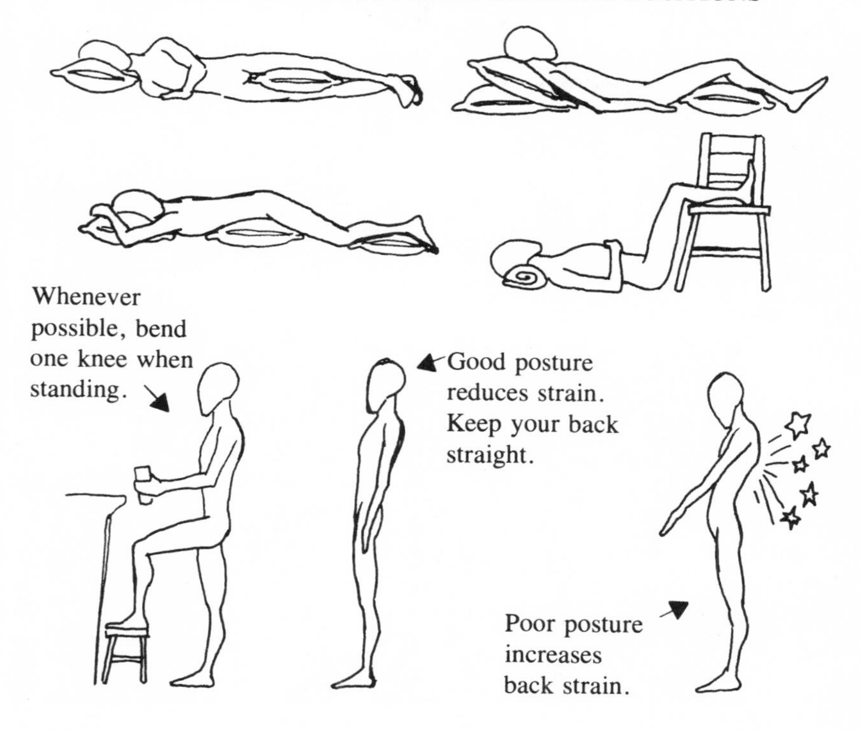

Try moving the car seat forward so that knees are bent.

Sit with knees higher than hips.

When reaching forward or overhead, keep your back straight and put one leg out behind you.

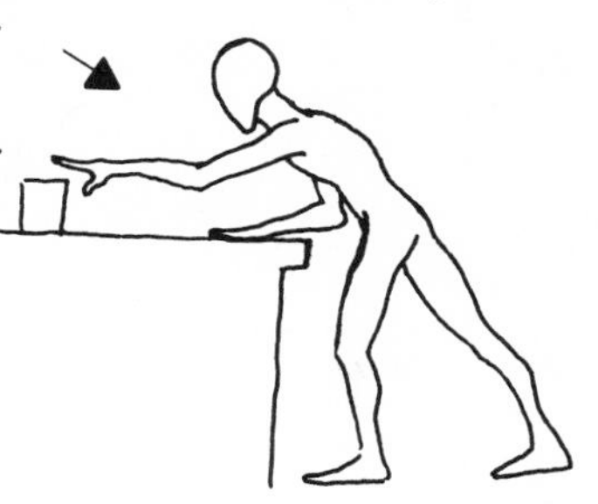

When carrying hold object close to you.

When lifting, bend hips and knees keeping back straight. Keep object close to you and lift with your legs.

Avoid lifting overhead by using a stool or ladder

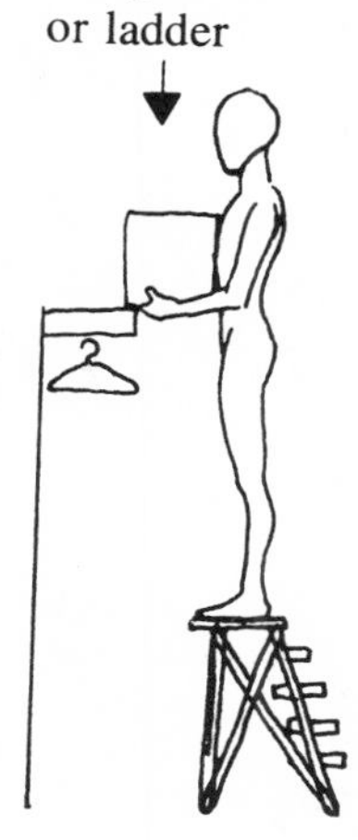

In daily activities involving bending or stooping, bend your knees, avoid twisting.

Home Treats Using Heat and Cold: (Check with Your Health Care Provider.)

Hot Moist Towels: Cover towels with plastic and wrap in a blanket. Apply to painful area for 15-20 minutes.

Commercial Hot Packs: Cover with several layers of dry towels. Apply for 10-15 minutes.

Tub Bath: Bath should be as warm as possible, for 10-20 minutes.

Strong Needle Power: Warm as possible for 5 minutes.

Electric Heating Pad: Set heat on LOW for 10-20 minutes. USE EXTREME CAUTION! Do NOT sleep all night with pad on—pressure on the pad can cause burns.

Moist Ice Towels: Fold wet towel in quarters and freeze in your freezer. Place in a pillowcase to prevent ice burn; apply for 10-20 minutes.

Ice Massage: Make ice in a small paper cup, rub ice directly over painful area for 5-15 minutes or until area becomes numb.

Ice Bag: Place ice bag on painful area for 10-20 minutes.

Commercial Cold Pack: Place in warm moist towel before applying to prevent ice burn. Apply 10-15 minutes.

By applying the illustrated suggestions you can decide what works best for your back problem. Make back care part of your daily routine to minimize strain and avoid possible injury. Consult with your doctor or your therapist regarding exercise, special activities, an increase in pain or a change of symptoms.

Adapted from OM-930 (2-74) GROUP HEALTH COOPERATIVE OF PUGET SOUND.

SUGGESTED READINGS FOR EXERCISE COURSE

The books, pamphlets, and articles listed here can increase your knowledge about fitness, health, and aging. Most of them will be available in public libraries. We encourage you to read those that interest you and to share what you learn with the rest of us.

Anderson, Bob. *Stretching.* P.O. Box 2734, Fullerton, CA 92633, 1975.

Basic Exercises for People Over Sixty, Moderate Exercises for People Over Sixty, National Association for Human Development, Washington, D.C.

Boston Women's Health Books. *Our Bodies/Ourselves.* NY: Simon & Schuster, 1971.

Christensen, Alice and David Rankin. *Easy Does It Yoga—For Older People*, The Light of Yoga Society, NY: Harper & Row Publishers.

Cooper, Kenneth, M. D. *The New Aerobics.* New York: Bantam Books, 1976.

Cooper, Mildred and Kenneth. *Aerobics for Women.* New York: Pocket Book Edition, 1969.

Devi, Indra. *Yoga for Americans.* Englewood Cliffs, NJ: Prentice-Hall, Inc., 1959.

The Fitness Challenge in the Later Years. DHEW Publication No. (OHD) 75-20802, Washington, D.C., May, 1968.
Frankl, Lawrence J. and Betty B. Richard. *Be Alive as Long as You Live.* Charleston, WV: Preventicare Pubs., 1977.
Fries, J. F. and Arapo, L. M. *Vitality and Aging: Implications of the Rectangular Curve.* Oxford, Great Britain: W. H. Freeman, 1981.
Harris, Raymond. *Guide to Fitness After Fifty.* New York: Plenum Publications Corporation, 1977.
Higdon, Hal. *Fitness After Forty.* Runners World, P.O. Box 336, Mt. View, CA 94042, 1977.
Hittleman, Richard. *Yoga for Personal Living.* Coronet, 1972.
Kraus, Hans. *Backache, Stress, and Tension.* New York: Pocket Book Edition, 1969.
Lande, Nathaniel. *Self-Health: The Lifelong Fitness Book.* NY: Holt, 1980.
Lindner, Erna Caplow. *Therapeutic Dance/Movement: Expressive Activities for Older Adults.* NY: Human Sciences Press, February 1979.
Orlick, Terry. *The Cooperative Sports and Games Book, Challenge Without Competition.* NY: Pantheon, 1978.
Rosenfeld. *New Views on Older Lives.* Prepared by NIMH Mental Health Science Reports Branch, Washington, D.C.
Senior Citizens Exercise Guide. The Travelers Insurance Co.
Shephard, Ray J. *Physical Activity and Aging.* Chicago: Year Book Medical Publications, Inc., 1978.
Smith, E. L. and Robert Sufass (eds.). *Exercise and Aging: The Scientific Basis.* Papers presented at the American College of Sports Medicine Annual Meeting, May 28-30, 1980, Las Vegas, Nevada. Enslow Publishers, Hillside, NJ 1981.
Sorenson, Jacki. *Aerobic Dancing for Physical Education.* KBH Productions, Inc., 1978. Distributed by: Educational Activities, Box 392, Freeport, NY 11520.
Zohman, Lenore. *Beyond Diet . . . Exercise Your Way to Fitness and Heart Health.* American Heart Association and the Presidents Council on Physical Fitness and Sports, CPC International, Inc., 1974.

Personal and Community Self-Help

These four class sessions cover a variety of topics—assertion training, communication skills, values clarification, and community issues. The participants in the Personal and Community Self-Help class should have the opportunity to generate their own lists of issues that they would like to address.

Once the topics are chosen, you can call upon a host of organizations and other community resources to help you deliver the program. For example, if the participants are interested in accident prevention, occupational therapists from local hospitals may be able to present ideas and information on how to accident-proof homes, especially for disabled people. Both your local fire and police departments have speakers that will be glad to talk with the participants. Any number of local organizations will be willing to send speakers, audiovisual materials, or conduct tours. The sky is the limit!

SESSION 1: AN OVERVIEW OF PERSONAL AND COMMUNITY SELF-HELP

Session Agenda

1. Welcome and introduce participants.
2. Preview the agenda for the session and make announcements.
3. Discuss the purpose and goals of the course.
4. Clarify the relationship between Personal and Community Self-help and the other three core content areas.
5. Define aggressiveness, nonassertiveness, and assertiveness.
6. Introduce the Affirmation of Health form and explain its purpose.
7. Evaluate the session.
8. Review the agenda for Session 2.

Handouts

Handout 1A – Definitions: Aggressiveness, Nonassertiveness, Assertiveness
Handout 1B – Affirmation of Health (sample)
Handout 1C – Affirmation of Health

Materials Needed

Flip chart
Magic markers
Session agenda printed on flip chart
Agenda for Session 2 printed on another page of the flip chart

Activity 1: WELCOME AND INTRODUCE PARTICIPANTS.

In order to develop a sense of familiarity and ease within the group, make use of an introductory ''game'' requiring some element of self-revelation. For example: Ask each person to introduce him/herself by telling the group something about his/her name: what it means, its ethnic origin, who he/she was named after. Facilitator may begin.

Activity 2: PREVIEW THE AGENDA FOR THE SESSION AND MAKE ANNOUNCEMENTS.

Methodology: Flip chart

Direct the group's attention to the printed agenda on the flip chart. Briefly list each item. Make any necessary announcements. Invite participants to make any announcements they may have.

Activity 3: DISCUSS THE PURPOSE AND GOALS OF THE COURSE.

Methodology: Lecturette

This course is designed to increase the participant's ability to assume responsibility for their personal health, as well as for the health of their community. More specifically, it will enable them to:

—become more aware of their values;

—learn how to make decisions that better reflect their values;
—express personal opinions and feelings without devaluing those of other people and exercise their rights without denying the rights of others;
—gain knowledge and skills to create a healthier and safer environment; and
—help each other make and maintain positive lifestyle changes in all four program areas.

Activity 4: CLARIFY THE RELATIONSHIP BETWEEN PERSONAL AND COMMUNITY SELF-HELP AND THE OTHER THREE PROGRAM AREAS.

Methodology: Directed group discussion

Ask participants what they perceive to be the relationship between Personal and Community Self-help and the other three program areas (stress management, nutrition, physical fitness).

Some points to consider include:

—The health of individuals (which the entire program seeks to enhance) is dependent in part upon a healthy, safe environment.
—Values clarification, assertiveness skills, and other communication skills practiced in the PCS class are useful in making and maintaining changes in nutrition, exercise, and in stress management;
—Becoming more involved in promoting the quality of one's life enhances personal health and wellbeing.

Activity 5: DEFINE "AGGRESSIVENESS," "NONASSERTIVENESS," AND "ASSERTIVENESS"

Methodology: Role play
Handout 1A

Refer to Handout 1A. Request three volunteers to read the definitions on the handout.

After all three definitions have been read, demonstrate each through a roleplay with a co-facilitator or assistant. Begin with aggressiveness, follow with nonassertiveness, and end with assertiveness. After each role play, encourage the group to express reactions and responses. Leading questions may be used to elicit discussion. For example, "Do you think the woman in the role play handled the situation appropriately?" "What do you think she could have done differently?"

Suggested role plays:

AGGRESSIVENESS

Scenario: An older woman at a grocery store checkout stand has just requested that her groceries be bagged in her cloth shopping bag. The clerk has refused to do so, stating that store policy requires her to bag the groceries in the standard paper bag.

Customer: (Shouting) "How absurd! You're all so unreasonable! Keep your old groceries then . . . see if I care." (She storms out of the store.)

NONASSERTIVENESS

Scenario: Same as above.

Customer: (Meekly, with eyes downcast) "Oh, well, O.K., then. I guess it's O.K."

Clerk: (Bags the groceries in the standard paper bag. She makes no comment.)

ASSERTIVENESS

Scenario: Same as above.

Customer: (Calmly, firmly). "I really would prefer to use this cloth shopping bag. I feel that the consistent use of new paper bags is wasteful and, in the long run, destructive to the environment."

Clerk: (Impatiently) "Madam, I really don't care. My supervisor would disapprove if I did not give you this regular paper grocery bag."

Customer: (Again firmly, calmly.) "I would like to speak with the manager then to explain that I really cannot accept the paper bag. Please call him or her for me."

Clerk. (With impatient sigh.) "Oh, here then, hand me your cloth bag." (She bags the groceries.)

Customer: "Thank you."

Activity 6: INTRODUCE THE AFFIRMATION OF HEALTH FORM AND EXPLAIN ITS PURPOSE.

Methodology: Handout 1B
Handout 1C

Ask participants to refer to the Affirmation of Health sample and to the blank Affirmation of Health (Handouts 1B and 1C).

Explain that at the end of every session each person will write down an activity that he/she wishes to accomplish in order to reach a personal goal. This goal is an affirmation of an agreement with oneself to enhance one's health. It may be an agreement to be more assertive or more environmentally active in a specific way.

Each person's efforts to accomplish the goal will be supported by other members of the class. The opportunity to develop this support will be the sharing period at the outset of every session. During sharing, individuals will be encouraged to talk with each other about their experiences in carrying out the agreed-upon activity. (Further elaboration on the sharing component is provided under PCS Session 2, Activity 2.) Read through the sample form with the group. Use it to explain the process of filling out their own forms.

Request that the group break into pairs or threesomes and, working together with those partners, fill out their individual forms for the first week. Each person should agree to have supportive contact (by telephone or in person) with a partner(s) during the coming week.

When this activity is completed, reconvene into the large group.

Activity 7: EVALUATE THE SESSION.

Methodology: Group discussion
Flip chart

The purpose of the evaluation is to continually improve the course by incorporating participants' feedback and perceptions. To elicit evaluatory comments ask the group, "What were the most useful aspects of this session?" and "How could the session have been improved?" It is highly important that the facilitator convey receptivity to constructive criticism.

Activity 8: REVIEW THE AGENDA FOR SESSION 2.

Methodology: Flip chart

The agenda for Session 2 should be written on the flip chart. Briefly read through it aloud.

HANDOUT 1A

DEFINITIONS: AGGRESSIVENESS, NONASSERTIVENESS, ASSERTIVENESS

Aggressiveness

You are behaving aggressively when you use hostile words or actions that force others to give in to your own preferences. As an aggressive person, you are interested in "winning," and you attempt to "win" by any means possible, including offending, hurting, or humiliating other people.

Although you may not feel especially anxious and you may "win," the price you pay can be very high. Because you violated the rights of other people, they will probably respond with dislike, hostility, and perhaps, act aggressively with you. You may find that others are beginning to avoid you and that they react to you sarcastically or hostilely. You may also often find yourself in arguments and left out of the social plans of others.

Nonassertiveness

When you do not express your own needs, allow your rights to be ignored, or let others make decisions for you that you would like to make yourself, you may be acting unassertively.

It is a wonderful thing to give selflessly to other people, but this giving should be something *you* want and choose to do, not something others tell you to do or something you think you should do to please others.

Nonassertiveness may cause you to value yourself less and to sometimes feel humiliated. You may feel helpless, controlled, and bitter because you rarely say or get what you want. It is often true that people feel self-hatred and resentment toward others when they are not expressing themselves assertively.

Assertiveness

You act assertively when you act in your own best interests, stand up for yourself, and express your opinions, feelings, and attitudes honestly, without devaluing others, and without feeling guilty or unduly nervous.

As an assertive person, you exercise your own rights without denying or violating the rights and feelings of others. Being assertive means you value yourself. It means stating your preferences in a way that causes others to take your ideas seriously.

Assertiveness does not guarantee that you will "win" in all situations, but it does make it easier to express yourself. Because you express yourself more easily, it is more likely that you will feel good about yourself, be satisfied with the outcome of a situation, and get what you need.

Adapted from Richey, C. A. *Workshop Manual for Trainers,* University of Washington, Seattle, 1979.

HANDOUT 1B

AFFIRMATION OF HEALTH

NAME Emily Rose

1. The major goal I want to accomplish in the next 4 weeks is:

Practice being more assertive!

2. How I plan to accomplish this goal is:

Week	Activity	Where	When	Times Per Week	With Whom
1	return shoes	store	this week	1	store clerk
2	be more affectionate	everywhere	this week	daily	friends & classmates
3	Ask a person not to smoke	restaurant	at dinner	1	smoker
4	smile more	In public	this week	daily	strangers

3. In order to help me follow through with my activity, I will:

______ Keep a journal of my thoughts, feelings, and reactions.

______ Keep records of "Daily Activities Log."

__✓__ Invite a friend or family member to work with me in my effort to change: the way this person will help me is: to take 5 minutes to talk with me and encourage me to be more assertive

__✓__ Invite someone from the class to work with me in my effort to change: the way this person will help me is to call me on the phone to see how I am doing on meeting my goals.

__✓__ Reward myself by taking two afternoon walks in the park.

HANDOUT 1C

AFFIRMATION OF HEALTH

NAME________________________

1. The major goal I want to accomplish in the next ______ weeks is:

2. How I plan to accomplish this goal is:

Week	Activity	Where	When	Times Per Week	With Whom

3. In order to help me follow through with my activity, I will:

   ______ Keep a journal of my thoughts, feelings, and reactions.

   ______ Keep records of "Daily Activities Log."

   ______ Invite a friend or family member to work with me in my effort to change: the way this person will help me is:

   ____

   ____

   ______ Invite someone from the class to work with me in my effort to change: the way this person will help me is

   ____

   ____

   ______ Reward myself by ______________________________

   ____

SESSION 2: BASIC STRATEGIES FOR INCORPORATING PERSONAL AND COMMUNITY SELF-HELP INTO LIFESTYLE

Session Agenda

1. Preview the agenda for the session and make announcements.
2. Facilitate sharing of the Affirmation of Health forms.
3. Define "values clarification."
4. Outline the values-clarification process.
5. Lead participants in identifying community issues of most concern to them and in deciding upon which of these should be addressed within the course.
6. Guide participants in filling out the Affirmation of Health form for the following week's activity.
7. Evaluate the session.
8. Preview the agenda for Session 3.

Handouts

Handout 2A – Possible Personal and Community Self-help Issues and Topics.

Materials Needed

Flip chart
Magic markers
Session agenda printed on flip chart.
Agenda for Session 3 printed on flip chart.
Twelve small slips of paper for each participant.

Activity 1: REVIEW THE AGENDA FOR THE SESSION AND MAKE ANNOUNCEMENTS.

Methodology: Flip chart

Same as Session 1, Activity 2.

Activity 2: FACILITATE SHARING OF THE AFFIRMATION OF HEALTH FORM.

Methodology: Group discussion

Encourage individuals to describe the goal-related activity which, during Session 1, they self-contracted to do. Discuss what they did to carry out this activity, and their self-reward.

Recognize all attempts regardless of success or failure. Change takes time. Immediate results are not always to be expected; the very process of establishing and seeking goals is valuable.

Encourage participants to recognize each others' goal-related activities; facilitators' modeling of this is imperative. Such peer support may also be developed by asking the group for suggestions on how they might assist each other in accomplishing their goals. For example: one man relates that he self-contracted in the Affirmation of Health to exercise assertive behavior by requesting his neighbor not to play loud rock-n-roll music after midnight on week nights. He further relates that he had found himself too embarrassed to actually do so. Another participant might assist him by offering to meet after class to role play the encounter.

Activity 3: DEFINE "VALUES CLARIFICATION."

Methodology: Lecturette

Values clarification is a process whereby each person becomes more aware of his/her values so that he/she is more able to act in accordance with these values.

The purpose of values clarification is to achieve harmony between values and behavior—between what one says and what one does—both in personal life and public life. Some may feel that values are relevant only to personal life. However, if individuals exercise their values only in their personal life, public policy decisions are left up to a very few people in positions of power whose values may be very different from those of their constituency.

To live with integrity means to enact one's values in both private and public concerns. Doing so requires two things: clarity of values and the assertiveness skills to stand up for them. To clarify values and to uphold them (especially in today's highly complex society where we are bombarded by innumerable external pressures) may be one of the most significant gains of a lifetime.

Activity 4: OUTLINE THE VALUES CLARIFICATION PROCESS.

Methodology: Lecturette
Group discussion

One way of becoming more aware of personal values is to apply the following seven questions to any decision one makes:

Choosing:
1. Did I choose freely?
2. Did I consider alternatives?
3. Did I consider the consequences of each alternative?

Prizing
4. Am I happy with my choice?
5. Am I willing to state my choice publicly?

Acting
6. Am I taking action on my choice?
7. Am I incorporating my choice into some pattern of my life?

Activity 5: LEAD THE PARTICIPANTS IN IDENTIFYING COMMUNITY ISSUES OF MOST CONCERN TO THEM, AND IN DECIDING WHICH OF THESE SHOULD BE ADDRESSED WITHIN THE COURSE.

Methodology: Structured group exercise

The following exercise enables people to identify community issues most important to them. Because it is based on values clarification, it provides a smooth transition to Activity 5.

A. Distribute Handout 2A. (The facilitator may want to add additional items to this list.)
B. Ask each person to circle twelve issues of most importance to them.
C. While they are thinking, hand out 12 small slips of paper to each person.
D. Request each person to write the 12 issues on the slips of paper and place all 12 slips of paper on the table in front of him/her.
E. Instruct each person to indicate with numbers the three issues of greatest interest.
F. Collect the slips of paper.

After class, tally the issues chosen. Use the results to determine what community and environmental issue(s) to talk about in Session 4. (*Note*: If the course will contain more than four sessions, the results of this tally may be brought to Session 4 and the participants can use it to choose a number of issues to deal with in subsequent sessions.)

Activity 6: GUIDE PARTICIPANTS IN FILLING OUT THE AFFIRMATION OF HEALTH FORM FOR THE FOLLOWING WEEK'S ACTIVITY.

Methodology: Small groups discussion

Invite the group to ask any questions regarding the use of the Affirmation of Health form. Encourage questions so that all participants understand the use of this form.

Request the group break into groups of 2 or 3 (not necessarily the same grouping as in Session 1) for completing the form. Suggest the self-contract for this week be to engage in an assertive behavior that reflects a strong personal value.

Activity 7: EVALUATE THE SESSION.

Methodology: Group discussion

As in the evaluation of Session 1, ask "What were the most useful aspects of the session?" and "How could the session have been improved?'' Facilitator's receptivity to constructive criticism is essential.

Activity 8: REVIEW THE AGENDA FOR SESSION 3.

Methodology: Flip chart

The agenda for Session 3 should be written on the flip chart. Briefly read it aloud.

HANDOUT 2A

POSSIBLE COMMUNITY ISSUES/TOPICS

Pollution

- Noise
- Fresh water
- Ocean
- Chemical
- Herbicides
  Pesticides
- Solid waste
  - Litter
  - Sewage
- Recycling

Food

- ''Junk'' Food
- World hunger
- World food distribution
- Gardening
- False/misleading advertising
- Eating lower on the food chain
- Food co-ops
- Food buying clubs
- Food banks

Energy

- Coal
- Solar
- Oil
- Nuclear
- Wind
- Natural gas
- Hydroelectric
- Home energy check

Social

- Crime prevention
- Fire prevention
- Care of elderly and infirm
- ''Bunco'' schemes

Land Use

- Crowding
- Development
- Farmlands vs. urban sprawl
- Housing
- Condominium conversion

''Wildcard''
Other

- CPR
- Violence on T.V.
- Social Security revisions
- Endangered species

SESSION 3: GIVING CONSTRUCTIVE CRITICISM

Session Agenda

1. Preview the agenda for the session and make announcements.
2. Facilitate sharing of Affirmation of Health forms.
3. Introduce the Assertiveness Bill of Rights.
4. Lead a group discussion on giving constructive criticism.
5. Present a model for giving constructive criticism.
6. Direct participants to fill out the Affirmation of Health for the following week's activity.
7. Evaluate the session.
8. Preview the agenda for Session 4.

Handouts

Handout 3A – Assertive Bill of Rights
Handout 3B – List of Situations in Which Assertiveness May Be Helpful

Materials Needed

Flip chart
Magic markers
Session agenda printed on flip chart
Agenda for Session 4 printed on flip chart

Activity 1: PREVIEW AGENDA FOR THE SESSION AND MAKE ANNOUNCEMENTS.

Methodology: Flip chart

Same as Session 2, Activity 1.

Activity 2: FACILITATE SHARING OF AFFIRMATION OF HEALTH FORMS.

Methodology: Group discussion

Same as Session 2, Activity 2.

Activity 3: INTRODUCE THE ASSERTIVENESS BILL OF RIGHTS.

Methodology: Group discussion
Handout 3A
Handout 3B

Distribute Handout 3A. Ask all participants to read it initially to themselves. Request that an individual read the first item aloud. Engage the group in discussion of this item. Stimulate discussion by asking such questions as:

"Do you feel that this is indeed a right?" "Why or why not?"

"Does this statement differ from your experiences or beliefs?"

"What is an example from your lives of a behavior reflecting this right?"

"If you do consider this a right, how able are you to exercise it?"

Follow each item with a brief discussion. The discussion should enable participants to define which of these items they perceive as rights. Equal respect must be indicated toward any participant who expresses the belief that one or more of the items is not a "right."

Distribute Handout 3B. Read the first two paragraphs aloud. Ask all participants to read the list to themselves and raise any questions they may have. Suggest that they keep this list for ideas in filling out their Affirmation of Health throughout the course.

Activity 4: LEAD A GROUP DISCUSSION ON GIVING CONSTRUCTIVE CRITICISM.

Methodology: Directed group discussion

As a result of this discussion, participants should:

—understand what is meant by constructive criticism;
—appreciate the value of constructive criticism;
—be able to give an example of constructive criticism;
—become aware of the degree to which they are able to give constructive criticism.

Directive questions may be used by the facilitator to stimulate discussion. For example:

"What is constructive criticism?"
"Is it useful?" "Why or why not?"

"Is it difficult to do?"
"What are some examples of constructive criticism you have given or received?"
"What are examples of unconstructive criticism?"
"Have you ever wanted to give constructive criticism but been unable to do so?"

Activity 5: PRESENT A MODEL FOR GIVING CONSTRUCTIVE CRITICISM

Methodology: Flip chart
Lecturette
Role play

Direct participants' attention to the flip chart which says:

Giving Constructive Criticism

When you ____________________________.

I feel ________________________________.

I would prefer ______________________.

This simple verbal framework is a method for giving constructive criticism. Call attention to the fact that it does not allow the critic to blame the other person for *making* him/her feel a certain way; it does not include the statement, "You *make* me feel ________."

Demonstrate the model by means of a role play. Request a volunteer to play the role of the person receiving criticism. The following is a possible role play.

Facilitator plays role of a man constructively criticizing his wife.

> "When you leave the house without saying 'goodbye' to me, I feel insignificant and unlovable. I would prefer you to say 'goodbye' when you go out."

Direct the group to break into groups of two and to role play the model in a situation that could actually occur in their lives.

Activity 6: DIRECT PARTICIPANTS TO COMPLETE THE AFFIRMATION OF HEALTH FORM FOR THE FOLLOWING WEEK'S ACTIVITY.

Methodology: Small groups
Affirmation of Health form

Instruct each group of two, formed for role playing, to join another group of two. Ask them to discuss their Affirmation of Health activity for the following week. Suggest that this activity be a specific situation in which they would like to give constructive criticism.

As always in the Affirmation of Health, each person's activity will be distinct from that of the others. The purpose of the small group discussion is not to decide upon a common activity but rather to aid each other in identifying a suitable activity, and to specify ways they will support each other in carrying out this activity during the week.

Facilitator should move from group to group watching for any participants still having difficulty using this tool. Such persons should be assisted individually by facilitator.

Activity 7: EVALUATE THE SESSION.

Methodology: Group discussion
Flip chart

Same as Session 2, Activity 7.

Activity 8: PREVIEW THE AGENDA FOR SESSION 4.

Methodology: Flip chart

Same as Session 2, Activity 8.

HANDOUT 3A

ASSERTIVE BILL OF RIGHTS*

1. I have the right to have and express my own feelings and opinions and to experience the consequences.

---

*Adapted from *The New Assertive Woman*, by L. Bloom, K. Coburn, and J. Perlman and *When I Say No, I Feel Guilty*, by M. Smith.

This Bill of Rights was devised by Cheryl A. Richey, School of Social Work, University of Washington, for use in assertiveness training groups.

2. I have the right to take responsibility for my own behavior, thoughts, and emotions.
3. I have the right to change my mind.
4. I have the right to make mistakes and to be responsible for them.
5. I have the right to decide what comes first in importance in my life.
6. I have the right to ask for what I want.
7. I have the right to decide not to assert myself.
8. I have the right to get what I pay for.
9. I have the right to ask for information from professionals.
10. I have the right to say "no," "I don't know," and "I don't understand."

I have the responsibility to recognize that everyone else has these same rights!

HANDOUT 3B

LIST OF SITUATIONS IN WHICH ASSERTIVENESS MAY BE HELPFUL

The following list identifies situations where assertiveness may be helpful. Many people experience discomfort or anxiety in handling these interpersonal situations.

In the classes you will learn how to handle these situations. We suggest you review the list and note the situations which are difficult for you. Not all the situations apply to everyone; the list covers a wide range of situations.

1. Turn down a request to borrow your car.
2. Compliment a friend.
3. Ask a favor of someone.
4. Resist sales pressure.
5. Apologize when you are at fault.
6. Turn down a request for a meeting or date.
7. Admit fear and request consideration.
8. Tell a person you are intimately involved with when he/she does or says something that bothers you.
9. Ask for a raise.
10. Admit ignorance in some areas.

11. Turn down a request to borrow money.
12. Ask personal questions.
13. Turn off a talkative friend.
14. Ask for constructive criticism.
15. Initiate a conversation with a stranger.
16. Compliment a person you are romantically involved with or interested in.
17. Request a meeting or a date with a person.
18. Your initial request for a meeting is turned down and you ask the person again at a later time.
19. Admit confusion about a point under discussion and ask for clarification.
20. Apply for a job.
21. Ask whether you have offended someone.
22. Tell someone that you like them.
23. Request expected service when such is not forthcoming, e.g., in a restaurant.
24. Discuss openly with the person his/her criticism of your behavior.
25. Return defective items, e.g., in a store or restaurant.
26. Express an opinion that differs from that of the person you are talking to.
27. Resist sexual overtures when you are not interested.
28. Tell the person when you feel he/she has done something that is unfair to you.
29. Accept a date.
30. Tell someone good news about yourself.
31. Resist pressure to drink.
32. Resist a significant person's unfair demands.
33. Quit a job.
34. Resist pressure to use drugs.
35. Discuss openly with the person his/her criticism of your work.
36. Request the return of borrowed item.
37. Receive compliments.
38. Continue to converse with someone who disagrees with you.
39. Tell a friend or someone with whom you work when he/she says or does something that bothers you.
40. Ask a person who is annoying you in a public situation to stop.

Adapted from Gambrill, E. D., and Richey, C. A. "An Assertion Inventory for Use in Assessment and Research," *Behavior Therapy*, 1975, *6*, 550-561.

SESSION 4: APPLYING ASSERTIVENESS SKILLS TO COMMUNITY ISSUES

Note to Facilitator

In this phase of the course, focus shifts from interpersonal assertiveness skills to the application of these skills in addressing community issues of concern to the participants. A number of conditions influence the effectiveness of this phase.

1. *The degree to which participants have come to value assertiveness and have begun to develop either their abilities to behave assertively or their motivation to develop these abilities.* If the entire course is composed of more than four sessions, the facilitator may choose to spend one or more additional sessions teaching other basic interpersonal assertiveness skills such as, "saying no," or "nonverbal communication."
2. *The degree to which participants understand the rationale underlying the shift in focus.* The facilitator may contribute to this understanding by a skillful presentation of Activity 3 in this session.
3. *The degree to which the topics addressed are indeed significant in the eyes of the participants.* To this end the topic-choosing exercise in Session 2, Activity 5 is invaluable, as is Activity 5 of this session in which the facilitator reviews the process. Again, if the course is to include sessions beyond Session 4, participants should be actively involved in selecting the topics to be addressed. Activity 4 in this session provides for this involvement. In order to insure participants' receptivity, the topics chosen by the facilitator for Session 4 should be those for which greatest interest was indicated in Session 2, and it should probably not be a highly volatile or threatening topic.
4. *The degree to which the topics are suitable for active involvement by participants.* The topics chosen should be issues which participants could act upon outside of the class setting if they choose to do so. If this phase consists of only one or two sessions, the topics could be issues for which legislative action would be appropriate. Skills in influencing legislative bodies are easily transferrable to other issues which individual participants may pursue following completion of the course.

Session Agenda

1. Review the agenda for the session and make announcements.
2. Facilitate sharing of the Affirmation of Health form.
3. Present the rationale for the transition from assertiveness skills on an interpersonal level to the application of assertiveness skills to community issues.
*4. Explain the ground rules for applying assertiveness skills to community issues.
5. Review the topic selection process begun in Session 2, Activity 5.
6. Introduce the community issue to be addressed in this session.
7. Discussion of the topic.
8. Present skills for becoming actively involved in the issue presented.
9. Direct participants to fill out the Affirmation of Health for the following week's activity.*
10. Evaluate the course.

Handouts

Handout 4A – Ideas for Personal and Community Self-help Sessions
Handout 4B – Tips for Contacting Legislators
Handout 4C – Helpful Hints on Writing Letters to Your Legislator

Materials Needed

Flip chart
Magic markers
Session agenda printed on the flip chart
Course evaluation forms (found in the Tips for Facilitators section)

Activity 1: PREVIEW THE AGENDA FOR THE SESSION AND MAKE ANNOUNCEMENTS.

---

*If this is the final session of the course, Activities 4 and 9 are not appropriate and should be deleted. Replacement activities are discussed in Notes to Facilitator.

Methodology: Flip chart

Same as Session 2 and 3, Activity 1.

Activity 2: FACILITATE SHARING OF THE AFFIRMATION OF HEALTH FORMS.

Methodology: Group discussion

Same as Session 2 and 3, Activity 2.

Activity 3: PRESENT THE RATIONALE FOR THE TRANSITION FROM ASSERTIVENESS SKILLS ON AN INTERPERSONAL LEVEL TO THE APPLICATION OF ASSERTIVENESS SKILLS IN COMMUNITY ISSUES.

Methodology: Directed group discussion
Lecturette

Facilitator may encourage group discussion by asking participants what they perceive to be the reason for the shift in focus from the interpersonal level to the community level. He/she should be prepared to explain the following concepts and should do so unless they are arrived at by the group in discussion.

Up to this point in the course we have learned:

—the importance of being aware of our values;
—how to become more aware of our values and their relative strengths;
—we have the right to live according to our values;
—living in accord with one's values requires assertive behavior at times;
—how to be assertive in interpersonal relationships;
—our values also determine how we respond to issues in society.

We have not yet discussed what is necessary in order to exercise our values in societal or public issues. This is the aim of this phase of the course. As mentioned earlier, our personal health is in part determined by the health and safety of our community. By learning how to enhance the health and safety of our community, we contribute to the health of ourselves and our families.

Activity 4: EXPLAIN THE GROUND RULES FOR APPLYING ASSERTIVENESS SKILLS TO COMMUNITY ISSUES. (This activity is applicable only if the course includes more than four sessions.)

Methodology: Lecturette
Handout 4A

Explain to the group the following ground rules:

1. The range of topics that may be discussed is unlimited, depending upon the group's interests.
2. As a group, participants will decide what issues to address and are invited to assist in locating community resources through which to address them. Examples of such resources are provided in Handout 4A.
3. One's ability to enact one's values in community issues requires three things: interest in the issue, some knowledge about the issue, and skills for influencing the issue. Therefore, with each topic we will seek to gain more knowledge about it and skills for affecting it.
4. We will strive to look at more than one side of any controversial topic.

Ask whether these ground rules are acceptable. Modify them if necessary.

Activity 5: REVIEW THE TOPIC SELECTION PROCESS BEGUN IN SESSION 2, ACTIVITY 5.

Methodology: Lecturette
Flip chart

Remind group of the topic selection exercise in Session 2, in which they identified three community issues of primary concern. Present on the flip chart the results of the tally. (Each item should be listed together with the number of people who had chosen it as one of the top 1, 2, or 3 items of interest to him/her.) Explain that you chose one of the highest scoring topics from that list for discussion in this week's session.

Activity 6: INTRODUCE THE COMMUNITY ISSUE (TOPIC) TO BE ADDRESSED IN THIS SESSION.

Methodology: Lecturette

Introduce the speaker or film. Remind the group that following the presentation they will discuss both the content of the presentation and ways to act upon the issue presented.

At this point some form of presentation on the topic is made (e.g., speaker, film, panel).

Example: If the topic for the session is crime prevention, a person from the local police department may be invited to speak. (Other resources and ideas are provided in Handout 4A.)

Activity 7: DISCUSSION OF THE TOPIC.

Methodology: Group discussion

Invite the group to ask questions and to express contrasting viewpoints; it is their right to disagree. Encourage them to share knowledge and experiences they have concerning the topic. If necessary, provide a model by questioning the speaker or expressing an opposing opinion.

If the presentation was highly biased, it is imperative to explain that the contrasting perspective could be presented in the following session if the group so chooses.

Activity 8: PRESENT SKILLS FOR BECOMING ACTIVELY INVOLVED IN THE ISSUES PRESENTED.

Methodology: Directed group discussion

The skills presented depend upon the issue. As indicated in this session's *Note to the Facilitator*, legislative advocacy skills are recommended because of their applicability to a wide range of issues. These skills are presented in this sample topic, crime prevention, especially with respect to older persons.

Example: Initiate a discussion of ways to become involved in the issue of crime prevention. List all suggestions on flip chart. Tap the experience of group members and of the speaker. Explain that time does not allow for in-depth study of each idea listed, and that in the remainder of the session, focus will be on one method—legislative advocacy. Legislative advocacy could be influential in developing or maintaining victim assistance programs, Neighborhood Watch programs, and the like.

Ask participants to list ways of influencing legislation. List all suggestions on the flip chart. Be sure to include:

—writing letters to legislators;
—using toll-free hotlines to call state legislators;
—sending telegrams;
—inviting legislators to public meetings;
—drawing up a petition and circulating it;
—developing an initiative.

Distribute Handout 4B. Ask participants to briefly list organizations that may be involved in legislative advocacy around that issue. Record the list on the flip chart. If no organizations are suggested, propose that before Session 5, each person identify and contact one and bring information (verbal or writen) about it to class.

Pass out Handout 4C. Request a volunteer to read it aloud.

Activity 9: DIRECT PARTICIPANTS TO FILL OUT THE AFFIRMATION OF HEALTH FOR THE FOLLOWING WEEK'S ACTIVITY.

Methodology: Affirmation of Health form

Propose to the participants that this week's activity be practicing their assertiveness skills in the arena of community issues, and building upon the knowledge shared in this session regarding the assertion of one's values through legislative advocacy. Suggest that each person choose one of the techniques discussed in this session and use it to influence a legislator with regard to either the issue of crime prevention or another issue of personal concern.

If this is the last session in the course, facilitator should strongly recommend that participants continue using the Affirmation of Health form on a regular basis to maintain the healthy changes they have made and to continue making positive lifestyle changes.

Activity 10: EVALUATE THE PERSONAL AND COMMUNITY SELF-HELP COURSE.

Methodology: Group discussion
Evaluation forms (available in the "Tips for Facilitators" section)

Explain the importance of the participants' honest and thorough evaluation to continually improve the course for future participants. Pass out course evaluation forms and request participants complete them before leaving.

HANDOUT 4A

IDEAS FOR PERSONAL AND COMMUNITY SELF-HELP

1. View a series of movies with environmental themes.
2. Crime prevention.
3. Tour of a recycling business.
4. Preparation of a meatless meal.
5. Visit to a grocery store and warehouse; visit to a food cooperative and one of its distributors.
6. Tour an arboretum or city park.
7. Nature hike to watch birds and identify plants, animals, and insects in their natural habitats.
8. Organize a food-buying club.
9. Tour of a bicycle co-op, or presentation by a cycling club.
10. Tour of and/or participation in a community garden program.
11. Tours of individual participants' gardens.
12. Have city utility do a home energy conservation inspection on a participant's home.
13. Presentation by a legislator on what they're doing in environmental areas.
14. Home safety.
15. Visit the local aquarium or natural history museum.

HANDOUT 4B

TIPS FOR CONTACTING LEGISLATORS

Mail campaigns are often used to influence politicians on important issues. The fate of a particular bill may hinge on a last minute outpouring of public sentiment which mail campaigns generate. To make *your* message count, here are a few tips on how to effectively contact your legislators.

Work together and present a unified voice. Keep a close watch on key issues and timing so a communications network can notify members when to contact their representatives. The best times to influence political action are when the bill in question is introduced or written during committee hearings on the bill, when the bill is sent to the floor, and when the bill is sent to the Governor or President for signing.

When group action is needed, a network or phone chain can be established where each member contacts six other members, who then call six more members, and so on. This method is quick and effective.

Be clear and to the point. Be sure to identify the bill or issue by its correct name. Try to give specific facts that the legislator can use later in arguing the case. Show how the bill will affect local economy, citizens, and the environment. Be specific and factual.

Send a written message. It provides documentation of your opinion. A call to the legislator leaves no such record. If time is short, you can send a mailgram or telegram quite cheaply and the charge will appear on your telephone bill. A personal opinion telegram will usually reach a representative within 24 hours at the state or national capitol offices. A 20-word message costs approximately $3.50, not including name, address, and telephone number. (There is no additional charge for this information.)

Special rates can be obtained to send identical messages to everyone in Congress at the same time. It costs $200 to reach all 100 senators and 435 representatives. The charge is $100 to senators only and $150 to representatives only.

Mailgram service allows you to send 50 words (including name, address, salutation, and closing) for $3.20 or 100 words for $4.10. These messages are transmitted to a post office near the addressee and sent in the next business day's mail. Be sure to include the zip code on the messages.

Developed by Washington Wilderness Coalition, Seattle, WA.

HANDOUT 4C

HELPFUL HINTS ON WRITING LETTERS TO YOUR LEGISLATORS

Do's

1. Determine what your issue is and stick to that *one* subject.
2. Identify the bill or issue in your letter. About 20,000 bills are introduced in each congress, so be specific. Try to get the bill number or describe it by its popular title.

3. Your letter should be timely. Write while there is still time to take action.
4. Be personable. Neatly handwritten or typed letters on your own stationery carry the most weight.
5. Write your *own* views, not those of someone else. Relate how the issue affects you personally.
6. Be constructive. If you find a problem, give suggestions for improvement.
7. Compliment your representative when deserved. In light of all of the complaints he or she receives, this would be much appreciated.

Don'ts

1. Don't make threats or promises.
2. Don't scold your congressperson or legislator.
3. Don't pretend to have vast political influence.
4. Don't write on every single issue that comes up.

Common Health Concerns

ACCIDENTAL INJURIES

Facts Relevant to Older Adults

—Accidental injuries are the seventh leading cause of death among those over 65.
—An estimated 24,000 to 28,000 elderly persons die yearly from accidents.
—Each year, another 800,000 persons 65 and over are injured badly enough to be disabled for at least one day.
—The most common causes of accidental injury among the elderly are falls, auto accidents (including pedestrian-auto accidents), and fires.
—Over 80% of deaths caused by falls occur among those over 65.
—Most falls occur in or around the home.
—33% of fatal burns occur among those over 65.
—26% of pedestrian fatalities occur among older adults.
—A common cause of accidental death among the elderly is susceptibility to excessive hot or cold external temperatures.
—Auto accidents are common among the elderly due to their often impaired senses and slower reflexes.

Risk Factors for Accidental Injuries among Older Adults

The physical changes that accompany aging such as impaired sensory perception, physical weakness, and impaired balance contribute to the increased risk of accidental injuries.

Aging also causes a slowed healing capacity and "brittle bones" due to osteoporosis.

Inability to tolerate extremely hot or cold external temperature is a factor in physical injury to older adults.

Poor lighting in the home, along with the impaired vision of older adults, increases the risk of accidental injury.

Stairs, waxed floors, wrinkled and slippery throw rugs, and spilled liquids on floors increase the risk of falls in the home.

Mental confusion and apathy due to depression, drug reaction, alcohol, or organic disease are risk factors for injuries in older adults.

Special Concerns for Older Adults

The social isolation of many older adults contributes significantly to their increased risk of accidental injuries and the more severe consequences of these injuries to this group. Because many people 65 and older are isolated from regular social contacts, if injured while alone it may be some time before the older adult is found and helped. Depression and apathy which can be brought on by social isolation also contribute to increased risk of accidental injury among the elderly.

Purpose of Health Promotion

Provide older adults with an increased awareness of the potentially serious consequences of accidental injuries due to falls, fire, and auto accidents, and with information that can help them to decrease their risk of accidental injuries in and around their homes, on city streets, and in their automobiles.

Topics for Health Promotion Classes and Activities

—Methods for making homes safer and "accident proof," especially for those with sensory losses
—Managing with extremely hot or cold temperatures
—Preventing and treating heat stroke and heat exhaustion
—Preventing and treating hypothermia (lowered body temperature from exposure to cold)
—First aid measures for accidental injuries
—Fire protection measures for the home (including smoke alarms)
—Special precautions for mobile home dwellers
—Precautions for the older adult driver
—Nonstrenuous exercises which help older adults keep fit and reduce their risk of serious injury in an accident

Resources for Education about Safety and Accident Prevention

Films, leaflets, pamphlets, textbooks, and courses on accident prevention and first aid.

National Fire Prevention Association
470 Atlantic Avenue
Boston, MA 02210

Pamphlets entitled *Fire Prevention All Over Your Home; Home Fire Detection; Clothing Can Burn!* and *Exit: Escape from Fire Wherever You Are.*

National Safety Council
Home Safety Information
425 North Michigan Avenue
Chicago, IL 60611

A ''Safety of the Elderly Kit,'' which includes pamphlets, worksheets, and checklists for community safety education programs.

National Retired Teacher's Association-American
Association of Retired Persons
1909 K Street, N.W.
Washington, DC 20049

NRTA-AARP also conducts a Driver Improvement Program to teach driving skills, publishes a booklet entitled *Retirement Safety Guide*, and has safety program packages for use in the community. The following 45-60 minute packages come complete with Program Leader's Manual, handouts, and a slide presentation:

1. ''Fire—Beware!'' presents fire hazards in the home and how to prevent them.
2. ''Falling—The Unexpected Trip'' helps people design a ''fall preventive'' home environment.
3. ''Safe Living in Your Mobile Home'' deals with the special safety problems of mobile homes.

A booklet of safety hints entitled *Handle Yourself With Care* (Pub. No. ODH–75–20805), and the accompanying guide, *Handle Yourself With Care: An Instructor's Guide for an Accident Prevention Course for Older Americans.*

U.S. Government Printing Office
Washington, DC 20402

First aid training and information are available at local chapters of the American Red Cross.

ALCOHOL ABUSE

Facts Relevant to Older Adults

—It is estimated that 2-10% of people over 60 years of age have an alcohol problem; as many as 1.6 million Americans over age 65 may have an alcohol problem.
—Older people in institutions, widowers, and people with medical problems have an even higher incidence of alcohol abuse.
—If a drinking problem began before a person was 40-50 years old, the treatment needs are the same as those of younger alcoholics.
—If a person has an alcohol problem that started late in life, treatment will need to focus on resolving the stresses that precipitated alcohol abuse.
—Alcohol can mask pain that is a symptom of disease and prevent the person from seeking treatment.
—Accidents caused by impaired judgment or coordination due to drinking are a health risk for older adults.
—Anyone can be an alcoholic—the problem is in detecting the drinking problem.
—Many older women have believed all their lives that alcohol use is unacceptable and may refuse to admit they have a drinking problem.

Health Risks for the Alcohol User

—Increased chance of organic disease
—Decreased cardiac efficiency
—Liver or pancreatic disease
—Cirrhosis of the liver
—Drug and alcohol interactions
—Increased frequency of medical problems
—Anemia
—Accidental injury
—Family history of alcohol abuse

Cues to a possible drinking problem in an older person are: self-neglect, falls, confusion, quarreling or other disruptions in family relationships, and an increase in medical problems.

Special Concerns for Older Adults

Stresses associated with aging, such as poor health, loss of status, bereavement of friends or relatives, are risk factors for alcohol abuse by older adults. In addition, aging diminishes physical reserves and increases the chance of organic disease that can be exacerbated by alcohol.

Purpose of Health Promotion

Provide relevant information on how alcohol affects older adults, the warning signs of a drinking problem, how to use alcohol without abuse and what to do if an older adult or friend has a drinking problem.

Topics for Health Promotion Classes and Activities

—Short- and long-term consequences of alcohol use
—Cautions about alcohol use for older adults
—How to drink without having a drinking problem
—Risk factors for a drinking problem
—How to gain access to community alcohol abuse programs
—How to deal with family or friends who have a drinking problem

Resources for Education about Alcohol Abuse

Alcoholics Anonymous (AA) sponsors local support groups for alcoholics and for their families. Membership is open to all; there are no dues. AA is supported by contributions. AA also publishes information about the purpose of AA and about alcoholism. "Time to Start Living" is a pamphlet specifically geared to the older alcoholic.

Alcoholics Anonymous World Services, Inc.
Box 459, Grand Central Station
New York, NY 10017

Information about alcohol, alcoholism, and treatment programs.

National Clearinghouse for Alcohol Information
Box 2345
Rockville, MD 20852

National Council on Alcoholism, Inc.
733 Third Avenue
New York, NY 10017

ARTHRITIS

Facts Relevant to Older Adults

—Arthritis means "inflammation of a joint" and is commonly used to describe over 100 different conditions which can cause aching, pain, swelling, and stiffness in joints or connective tissue.
—Arthritis affects one in seven Americans and claims one million new victims every year.
—Studies have shown that 97% of individuals over the age of 60 have enough joint degeneration to be apparent on x-rays, although many of these people have no symptoms.
—Arthritis is usually a chronic and progressive disease that has no cure; but there are techniques of treatment and management of arthritis that can significantly slow the progress of the disease and significantly reduce the pain and crippling it can cause.
—Arthritis can cause severe pain as well as lead to deformity and crippling if untreated.

Health Practices that Help Control the Impact of Arthritis on Lifestyle

—Medication
—Exercise and rest
—Weight Reduction
—Heat
—Surgery
—Rehabilitation for disabilities due to arthritis

Special Concerns for Older Adults

Older people who develop arthritis may assume it is an inevitable part of aging and not seek help from a health care provider.

Purpose of Health Promotion

Teach older adults to recognize the warning signs of arthritis and to practice self-care skills that help control arthritis such as relaxation and exercise techniques; provide information on different types of arthritis and management programs, and encourage adoption of health practices that will help prevent arthritis.

Topics for Health Promotion Classes and Activities

—Warning signs of arthritis
—Varieties of arthritis—description and risk factors
—Arthritis drug and surgical treatments
—Myths and facts of arthritis treatment
—Demonstration of exercise and joint protection procedures for the arthritic person
—Providing (and getting) support for people with arthritis

Resources for Arthritis

Clearinghouse of information on materials and resources about arthritis.

Arthritis Information Clearinghouse
P.O. Box 34427
Bethesda, MD 20034

Conducts research and training; provides public information, education, and community services through local chapters. Publishes *Guidelines for Conducting Patient Education Classes* (for physicians), guidelines for establishing patient self-help groups, and a *Self-Help Manual for Arthritis Patients.*

The Arthritis Foundation
National Office
3400 Peachtree Road N.E.
Atlanta, GA 30326

Materials, resources, and audiovisual materials on arthritis.

Learning Resources Facility
Institute of Rehabilitation Medicine
400 East 34th Street
New York, NY 10016

Films and slides on arthritis.

Kaiser Permanente Health Center
Audiovisual Workshop
3799 Peidmont Avenue
Oakland, CA 94611

Training for instructors of arthritis self-management courses.

Stanford Arthritis Center
701 Welch Road, Suite 2208
Palo Alto, CA 94304

CANCER

Facts Relevant to Older Adults

—Cancer is the disease most feared by older adults.
—The incidence of cancer rises sharply until ages 85 to 90.
—Cancer is the second leading cause of death in the U.S. for older adults; over 60% of all cancer occurs among those over age 60.
—Skin and prostate cancers are primarily associated with old age.
—Cancer of the breast has a better prognosis in older persons than in younger persons; cancer of the thyroid and malignant melanoma have a worse prognosis for older adults.
—Older people have more misperceptions and myths about cancer than any other disease.
—Cancer in older persons is diagnosed at a later stage of disease than cancer in younger persons.
—Prejudice and discrimination against older adults negatively affects the motivation and abilities of professionals, older adults, and others to prevent and cope with cancer.

Risk Factors for Cancer Among Older Adults

Major known risk factors:

—Cigarette smoking
—Alcohol abuse
—Personal or family history of certain cancers
—Exposure to certain chemicals
—Excessive exposure to sun

Highly suspected risk factors:

—High fat diet
—High salt diet
—High levels of stress

Special Concerns for Older Adults

Older adults especially need to know that it is not too late to prevent cancer nor to limit the effects of cancer. Because of the confusion that surrounds cancer, older adults need accurate information to dispel the myths and fears about cancer in order to reach the highest quality of life possible.

Purpose of Health Promotion

Teach older persons to recognize the warning signs of cancer, supply information on different types of cancer, and provide opportunities to practice self-care skills that may help control cancer and reduce risk.

Topics for Health Promotion Classes and Activities

—Warning signs of cancer
—Myths about cancer
—Assessing personal risk for developing cancer
—Cigarette smoking and cancer
—Environmental causes of cancer
—Coping with cancer
—Information on different cancers
—Stress management and visualization for cancer control
—Cancer screening and detection programs
—Hospices

Resources for Education about Cancer

Films, pamphlets, and posters about cancer and coping with cancer are available from local chapter offices or the national office.

American Cancer Society
777 Third Avenue
New York, NY 10017

Provides written information about different kinds of cancer, treatments, and the effect of nutrition on cancer.

Office of Cancer Communications
National Cancer Institute
National Institutes of Health
Bethesda, MD 20014

DIABETES

Facts Relevant to Older Adults

—Diet is central to the treatment of diabetes.
—Diabetes causes circulation problems which can affect the legs and feet.
—Eighty percent (80%) of diabetics are over 45; 5% of the older population have diabetes and are being treated.
—Diabetics face an increased risk of developing heart and blood vessel disease.
—In older ages, rates of diabetes are higher in women.
—For the older adult with diabetes, the foot is often where the first overt complications of diabetes arise.
—Prevention offers the only logical approach to managing diabetes, particularly to avoid foot problems.
—Atherosclerosis progresses twice as rapidly in diabetics than in nondiabetics.
—Coronary artery disease is five times more common in the diabetic woman, twice as common in the diabetic man.
—Diabetics do not have a larger number of foot problems than the rest of the population, but their complications are more severe because of decreased blood supply and lessened sensation.
—Most adult diabetics are overweight.
—Diabetes is made worse by stress.

Risk Factors for Diabetes in Older Adults

—Overweight
—Family history of diabetes
—Stress
—Infection
—Surgical Procedures

Special Concerns for Older Adults

Older adults with diabetes need to check their feet daily for injuries because they often do not feel pain as easily as nondiabetics. An older adult diabetic may have one or more chronic diseases in addition to diabetes and may be taking more than one drug at any given time.

Purpose of Health Promotion

Provide older adults with information on the importance of diet and exercise in preventing or controlling diabetes as well as provide general information about diabetes and the older adult.

Topics for Health Promotion Classes and Activities

—Diabetes screening
—Food Substitutions
—Questions to ask your doctor about diabetes
—Hygiene and care of skin, feet, teeth, and gums
—Information on diabetes self-management
—Information on diabetes self-help groups

Resources for Education about Diabetes

American Association of Diabetes Educators
North Woodbury Road, Box 56
Pitman, NJ 08071

American Diabetes Association
600 Fifth Avenue
New York, NY 10020

Free bimonthly newspaper for diabetics.

Diabetes in the News
233 East Erie Street, Suite 712
Chicago, IL 60611

Newsletters, tours, and other special services for travel-oriented diabetics.

Diabetes Travel Services, Inc.
6 East 36th Street
New York, NY 10016

Information about foot care in a publication entitled *Feet First.*

Superintendent of Documents
Government Printing Office
Washington, DC 20402

Diabetes—New Look at an Old Problem has a specific section on diabetes and the elderly.

Lowenstein, Bertrand E. and Paul D. Preger, Jr. New York: Harper and Row, 1976.

Charts explaining diabetic diets, vegetarian diets, weight control, and fitness. A brochure is available.

National Health Systems
Box 1501
Ann Arbor, MI 48106

A 25-minute educational film, ''Live With It'' is available to help the newly-diagnosed diabetic learn to live with his/her illness. Available for purchase or rent on videocassette or 16mm film.

Teach'em, Inc.
625 North Michigan Avenue
Chicago, IL 60611

FOOT CARE

Facts Relevant to Older Adults

—Decades of weight-bearing can have negative effects on the aging foot.
—Changes with aging in the blood vessels of the feet and legs may lead to a loss of adequate blood supply to the feet.

—The skin of the foot becomes thinner and more delicate with age and cracks or tears easily, increasing the risk of infection which can start in cuts and scratches.
—People with diabetes or peripheral vascular disease run a significant risk of a foot infection that can lead to gangrene and amputation.
—With treatment and good foot care habits, much can be done to ease foot discomfort and maintain activity.

Risk Factors for Foot Problems among Older Adults

—Aging
—Poor footwear
—Lack of exercise
—Systemic disease (diabetes, rheumatoid arthritis, gout)
—Tight clothing
—Irregular foot care
—Lack of rest
—Cigarette smoking
—Untreated cuts, calluses, corns, bunions, warts, etc.
—Extreme temperatures on the feet

Special Concerns for Older Adults

Older adults need to know the basics of good foot care in order to alleviate or prevent foot problems that may limit mobility.

Purpose of Health Promotion

Teach older adults good foot care, relate healthy feet to larger issues of independence and self-care, and provide support for daily attention to foot care.

Topics for Health Promotion Classes and Activities

—Demonstration and practice of foot care
—Relaxation, exercise, and healthy feet
—Resources for foot care
—Foot care for the person with systemic disease
—Making foot care a daily habit

Resources for Foot Care

Pamphlets and articles on foot health, diabetes, arthritis, sports, and occupational hazards.

American Podiatry Association
20 Chevy Chase Circle, N.W.
Washington, DC 20015

Materials on Buerger-Allen exercises for the feet.

Canadian Diabetes Association
123 Edward Street, Suite 601
Toronto, Ontario MFG 1F2 Canada

Diabetes Education
Metropolitan Medical Center
900 S. 8th Street
Minneapolis, MN 55404

Public Affairs Pamphlets
381 Park Avenue South
New York, NY 10016

(Pamphlet #345A, Light on Young Feet)

Check with local community clinics, Health Maintenance Organizations, Visiting Nurse Service, hospitals, and podiatrists about outpatient geriatric foot care clinics and resource people for teaching about foot care.

HEART CONDITIONS

Facts Relevant to Older Adults

—Heart disease is the leading cause of death of older adults in the United States.
—It is estimated that some degree of heart disease is present in almost everyone over 70 years of age.
—Lifestyle is a stronger factor than heredity in causing or preventing heart trouble.
—Behavioral change is the most effective treatment for heart disease, even for those who have already had a heart attack.

Risk Factors for Heart Conditions among Older Adults

—Amount of cholesterol and saturated fat in the diet
—High blood pressure
—Cigarette smoking
—Stress
—Heredity/occurrence of heart conditions in the family
—Excess weight
—Diabetes
—Lack of exercise

Special Concerns for Older Adults

Older adults need information, skills, and support to make lifestyle changes which help prevent development of a heart condition or help cope with a heart condition.

Purpose of Health Promotion

Teach older adults to evaluate individual risk levels for heart conditions, provide information on dietary, exercise, and personal habits that decrease risk, practice behaviors that will reduce risk, and provide group support for maintaining healthier lifestyles.

Topics for Health Promotion Classes and Activities

—Information on different heart conditions
—Assessing personal risk for heart attacks
—Aerobic exercise and the heart
—Reducing sources of cholesterol and saturated fat in the diet
—Cigarette smoking and the heart
—Stress reduction
—Blood pressure monitoring
—Warning signs of a heart attack
—Weight reduction and the heart
—What to do in case of a heart attack
—How to do cardiopulmonary resuscitation (CPR)
—Coping with a heart condition
—Health habits for a healthy heart

Resources for Heart Conditions

Information about cardiovascular disease and coping with heart conditions. Contact local chapters.

American Heart Association
7320 Greenville Avenue
Dallas, TX 75231

Provides copies of *Living With a Heart Ailment.* Ask for Pamphlet No. 521.

Public Affairs Pamphlets
381 Park Avenue South
New York, NY 10016

Offers a 1-hour community education package on cardiovascular health. Includes instructions and script, a short film, *Without Warning*, handout materials, and provides a guide for finding a local resource person.

National Retired Teachers Association/
American Association of Retired Persons
1909 K. Street N.W.
Washington, DC 20049

Information on health practices for healthy old age.

Life Extension Institute
1185 Avenue of the Americas
New York, NY 10036

HYPERTENSION

Facts Relevant to Older Adults

—Hypertension usually causes no symptoms until damage to the brain, heart, or kidneys has already occurred.
—It is estimated that over half the people in the United States who have high blood pressure are unaware of it.
—The incidence of hypertension increases in older age groups.
—It is estimated that nearly 40% of whites and over 50% of blacks over 65 have high blood pressure.
—Women are slightly more likely to have high blood pressure.
—Hypertension causes or contributes to 1,000,000 deaths annually in the U.S.
—Even slightly elevated blood pressure means a considerably greater risk of death due to heart failure, stroke, or kidney failure.

—In 90-95% of hypertension cases ("essential" or "primary" hypertension), the cause is unknown and it cannot be cured.
—Although the cause is not known, it is known that following certain health practices will considerably lower the risk of hypertension or hypertension-related disease and these same practices will help control hypertension already present.
—It is estimated that between ½ and ¾ of all diagnosed hypertensives do not comply with treatment programs.

Health Practices that Reduce Risk of or Control Hypertension

—Refrain from cigarette smoking
—Eat low-fat, low cholesterol foods
—Exercise regularly
—Take medication as prescribed
—Restrict salt intake
—Keep weight within normal limits
—Reduce stress
—Family history of hypertension

Special Concerns for Older Adults

Because hypertension increases in incidence among older age groups and often has no symptoms, older adults need self-care skills in monitoring blood pressure and developing health practices to prevent or control occurrence. For those with hypertension, the need for continued treatment should be emphasized.

Purpose of Health Promotion

Help older adults develop lifestyles that will reduce their individual risk for hypertension or control hypertension which has already developed.

Topics for Health Promotion Classes and Activities

—Facts and myths about blood pressure and hypertension
—Assessing individual risk for hypertension
—Diet for a healthy heart
—How to take own pulse and blood pressure
—Treatment for hypertension

—How to decrease salt and cholesterol in the diet
—Practice preparation of low-salt, low-fat food

Resources for Hypertension

Printed materials, films, and slides on hypertension. Also keeps a roster of speakers.

American Heart Association
44 East 235th Street
New York, NY

CORE Communication in Health, Inc.
919 Third Avenue
New York, NY 10022

National High Blood Pressure Education Program
National Heart, Lung, and Blood Institute
Bethesda, MD 20205

Many community clinics, health maintenance organizations, and hospitals offer patient education programs on hypertension. Blood pressure screening is often available at local health fairs.

INSOMNIA

Facts Relevant to Older Adults

—As people grow older, the amount and quality of sleep changes.
—Individual sleep patterns are unique.
—Older adults tend to sleep less, and less soundly, than when they were younger.
—Older women tend to report more sleep problems than older men.
—People over 64 consumed 33% of all prescription sleeping pills in 1974.
—Sleeping pills are usually tested on young, healthy adults and physicians may not be aware of the special risks to the older adult.
—Pain caused by body movements during sleep is one of the most common reasons older adults awaken during the night.

—Some conditions, such as arthritis, may be especially disruptive at night.
—Normal, healthy 80-year-olds spend 1/5 of the night awake.

Risk Factors for Sleep Problems

—Bad sleep habits
—Stress and anxiety
—Use of stimulants and other drugs
—Aches and pains

Special Concerns of Older Adults

Older adults are at a greater risk for sleep problems than other age groups. Physical problems or anxiety, fear, depression, even anticipation can keep any person awake but the older adult may be more sensitive.

Purpose of Health Promotion

To give an understanding of sleep and insomnia so that the older adult may be reassured or be alerted to seek help; to teach techniques to help falling asleep easier.

Topics for Health Promotion Classes and Activities

—How sleep changes throughout life
—An explanation of sleep and sleep cycles
—Factors that can either promote or interfere with sleep
—How drugs can affect sleep
—Stress management techniques

Resources for Education about Insomnia

Books that have information on sleep and the older adult.

Glickman, Stephanie and Judy Lipshutz. *Your health and aging.* New York: Division of Gerontology, Office of Urban Affairs Health Affairs, NYU Medical Center, 1981.

Hales, Dianna. *The complete book of sleep.* Reading, MA: Addison-Wesley, 1981.

MEDICINE MANAGEMENT

Facts Relevant to Older Adults

—Physical changes due to aging affect how a person's body handles drugs—older people are more prone to side effects and to adverse drug reactions than are younger people.
—Deliberate abuse of drugs such as barbiturates, analgesics (pain-killers), and bromides is common among persons over 65.
—Abuse of such over-the-counter drugs as laxatives, vitamins, aspirin, and cold medicines is a particular problem among older people.
—Self-medicating with drugs prescribed for past illnesses is dangerous.
—Drug-sharing is a particular problem in nursing homes and housing complexes for older adults.
—Drug misuse can result when drugs combine to lower effectiveness of one or more of the drugs taken, or when drug combinations have a toxic effect.
—Many people do not take medicines as prescribed.
—Health providers do not always check to make sure a patient understands how to take a medicine.
—Drug products may be poorly labeled.
—Advertising that solicits use of certain drugs by older consumers may lead to drug abuse.

Health Practices that Reduce the Risk of Medicine Mismanagement

—Maintain general health history records.
—Keep records of all drugs currently used, including prescription and nonprescription drugs.
—Monitor current symptoms.
—Record any past drug reactions.
—Share all relevant information with the health care provider including any physical, financial, or social conditions that might affect use of any drug prescribed.

Request the following information for any drug prescribed:

—What is the name of the drug?

—What was it prescribed for?
—How will it affect symptoms?
—What side effects could the drug have?
—What do I do if I have side effects?
—When should I take the drug?
—How long should I take the drug?
—Should I take it on an empty stomach, or with food, water, or milk?
—What substances should I avoid while taking the drug?
—How should I store the medicine?
—How much will the medication cost?
—Is there a less expensive, generic form available?

Special Concerns for Older Adults

For people who are taking more than one medication, as many older people do, the risk of drug mismanagement increases. Inadvertent misuse of drugs is the greatest drug-related problem of older people and reflects many of the other kinds of problems that beset older persons: low income, social isolation, physical impairments, physical changes due to aging, medical problems, and sometimes mental confusion.

Purpose of Health Promotion

Provide information on medicine management, training, and practice in discussing medication with a health care provider, pointers on keeping a personal medical history, and ways to self-monitor medicine consumption for older adults.

Topics for Health Promotion Classes and Activities

—How to talk to your doctor and what to ask
—Assertiveness training for health care problems
—How to keep a personal health history
—Memory aids for scheduling medication
—How to use over-the-counter medication
—The role of prescribed medicines in health
—Drug reactions and interactions
—Drug information resources in the community
—Drug and alcohol interactions
—How to use your pharmacist

Resources for Medicine Management

Offers an excellent package on medicine management called *Elder-Ed: Using Medicines Wisely.* Also offers a package for training service providers about "Wise Drug Use for the Elderly."

NRTA/AARP
1909 K. Street N.W.
Washington, DC 20049

Public Affairs Pamphlets
381 Park Avenue South
New York, NY 10016

Publishes pamphlet #570, "Know Your Medication: How to Use Over-The-Counter and Prescription Drugs."

Elder Ed
School of Pharmacy
University of Maryland at Baltimore
636 W. Lombard Street
Baltimore, MD 21201

Offers a video/film and patient education package, "Wise Use of Drugs—A Program for Older Americans," aimed at older consumers. No charge to borrow the film.

RHR Filmedia
1212 Avenue of the Americas
New York, NY 10035

Rents a film entitled "It's Up to You" on medicine management, using the pharmacist as a resource.

National Pharmaceutical Council
1030 15th Street N.W.
Washington, DC 20005

Consumer Protection education pamphlets:

#76-3006 "We Want You to Know About Labels on Medicines"
#74-3011, "We Want You to Know About Prescription Drugs"
#74-3005, "We Want You to Know About Adverse Reactions to Medicines"

Food and Drug Administration
Office of Consumer Affairs
5600 Fishers Lane
Rockville, MD 20852

NORMAL CHANGES OF AGING

Facts Relevant to Older Adults

—Although there are many individual differences, in general the human life span is more or less predictable.
—Physical changes that are fundamental to aging do not occur in isolation; they also have psychological and social significance.
—Aging is often defined in terms of the physical changes observed by both the older adult and by those around him/her.
—Normal changes are gradual and progressive, universal and intrinsic.
—After age 30, a wide range of physiological functions decline at a little less than 1% a year.
—An equally large number of physiological functions remain fairly stable until the older adult is well past the seventh decade.
—Support continues to increase for the idea that certain health practices slow biological aging, promote physical health, and that these activities are interdependent.

Special Concerns for Older Adults

People think of the physical changes of aging in terms of losses, and aging, therefore, is frequently viewed as an overwhelming negative experience.

Purpose of Health Promotion

Provide older adults with information about normal age changes and teach the health practices that promote health.

Topics for Health Promotion Classes and Activities

—Myths and stereotypes of aging
—Physiological changes of aging

—Organic Brain Syndrome/Senility
—Current theories of aging

Resources for Education on Normal Changes of Aging

Local chapters or the national organizations may have pamphlets, audiovisuals.

NRTA/AARP
1909 K. Street N.W.
Washington, DC 20049

Free publication, *Age Page*, covers concerns of older adults including age changes. Other information is available from them.

National Institute on Aging
Box MSC, Bldg. 31
Room 5C35
Bethesda, MD 20205

A book that has an easy reading style on aging and its changes:

Comfort, Alex. *A good age.* New York: Mitchell Beazley Pub., Ltd., 1976.

Local colleges/universities with aging programs in departments of Sociology, Social Work, or Nursing.

Institutes of Aging

City/County/State offices on aging

ORAL HEALTH

Facts Relevant to Older Adults

—Gum and mouth diseases increase with age, especially for smokers.
—Half of the people over 65 in the United States have no teeth; another 15% have lost either all their upper or all their lower teeth.
—Fixed incomes and restrictions in insurance or Medicare coverage prevent people from seeking dental care as often as they need it.

—People who need dentures or who need their dentures refitted comprise 40% of the over-65 population.
—Prevention of plaque buildup and gum disease is as important for people who have lost their teeth as for those who retain their teeth.
—Correct care of dentures prevents many oral problems.
—Regular dental checkups are vital for everyone.

Special Concerns for Older Adults

Loss of teeth as well as denture problems can lead to problems with chewing and, if the older person avoids all but a few foods that are easy to chew, malnutrition may result.

Purpose of Health Promotion

Teach older adults good oral health practices and provide information on behaviors that may increase the risk of oral health problems such as cigarette smoking, drinking lemon juice in warm water to aid digestion, and clenching pipe stems.

Topics for Health Promotion Classes and Activities

—A field trip to a dental clinic that provides dental services to older adults
—Demonstration of denture care and/or of teeth and gum care
—Demonstration of health aids for the disabled
—Suggestions for making oral health care a daily habit
—Myths and facts of good oral health care

Resources for Oral Health

Printed and audiovisual materials on dental health, including a film, "Options—Dental Health in the Later Years," for older adults that emphasizes that dental disease is not an inevitable result of aging. For rental/purchase.

American Dental Association
211 East Chicago Avenue
Chicago, IL

Pamphlet with large print for older consumers entitled "Caring for Dentures" and "Caring for Natural Teeth."

Department of Community Dentistry
University of Washington
Seattle, WA 98195

SENSORY LOSS

Facts Relevant to Older Adults

—Sensory losses and changes in older adults are gradual and rarely occur suddenly.
—Multiple losses (visual and auditory, particularly) are more difficult to cope with than single losses.
—The prevalence of hearing and vision loss is greatest in those over 65.
—About 30% of all older adults have significant hearing losses; men experience a greater loss than women.
—The hard of hearing differ from the deaf.
—Although most older adults need glasses, 80% have fair to adequate vision; women are more likely to have vision handicaps.
—Vision problems are the second most relevant chronic health problem and tend to change most rapidly in the elderly and require regular attention.
—Glaucoma, cataracts, and macular degeneration are eye diseases that can significantly affect vision. Sometimes the person is unaware of the changes until serious loss has occurred.
—Except for total blindness, there are successful interventions that usually restore significant vision. Even among the legally blind, 67% may have sufficient vision restored for important life functions.
—The interrelatedness of smell, taste, and sight results in the loss of great satisfaction and potential compensations for older adults.

Types of Sensory Loss

Vision: Reduced ability to see details up close
Spots in front of eyes
Dry eyes

	Watery eyes Cataracts Glaucoma Diabetic retinopathy
Hearing:	Conductive hearing loss Sensori-neural hearing loss Central auditory processing loss
Touch:	Failure to detect pain No response to being touched
Smell:	Nasal congestion No response to odors
Taste:	Ability to taste sweet, salt, bitter, and sour declines Change in taste of foods

Possible Health Consequences of Sensory Loss for Older Adults

—Depression
—Social isolation
—Loss of appetite
—Decreased mobility
—Failure to detect pain
—Permanent sensory loss
—Loss of independence
—Increase in accidents

Special Concerns for Older Adults

Older adults may associate sensory loss with stigma and increased dependence on others and therefore deny the loss and not seek information or treatment. Vision loss may be considered normal by the older adult who then does not seek care to restore sight.

Purpose of Health Promotion

Teach older adults to monitor their sensory abilities, recognize cues of sensory loss, adopt habits that prevent sensory loss, and recognize sources of treatment in their community. For vision, teach older adults to distinguish between normal and abnormal changes in vision, how to modify their environment to maximize vision performance, how and when to use visual aids.

Topics for Health Promotion Classes and Activities

—Signs of sensory loss
—Aids and compensation techniques
—Visual problems and resources
—Hearing problems and resources
—Finding medical care for sensory problems.
—Demonstration of the physiological measures of sensory capability
—Stress reduction, nutrition, and exercise relating to eye problems
—Coping with a sensory reduction or loss
—Information on and using community screening programs for hearing and vision

Resources for Sensory Impairments

Books, magazines, films, special aids, research, consultation, education, rehabilitation, legislative action services on vision problems. Contact state or regional offices for information.

American Foundation for the Blind
15 West 16th Street
New York, NY 10011

Materials on vision.

Auxiliary to the American Optometric Association
243 North Lindbergh
St. Louis, MO 63141

Materials on blindness and diabetes.

American Diabetes Association
1 West 48th Street
New York, NY 10020

Pamphlets in large print on hearing problems and hearing aids.

American Speech and Hearing Association
10801 Rockville Pike
Rockville, MD 20857

Deafness Research Foundation
366 Madison Avenue, Suite 705
New York, NY 10017

Environment Communication Intervention
for the Aging Project
Division of Communication Disorders
Department of Special Education
Murray State University
Murray, KY 42071

TOBACCO SMOKING

Facts Relevant to Older Adults

—Approximately 24% of older American men and 10% of older American women now smoke tobacco.
—More older persons were smoking in 1975 than in 1970 despite widespread publicity about the health risks incurred by smoking.
—Smoking cessation results in health benefits for people of any age.
—Cancer is the second leading cause of death of people over 65 and one out of every five fatal cases of cancer is cigarette related.

Health Risks of Tobacco Smoking

—Coronary heart disease
—Increased blood pressure
—Cancer of the lung and other organs
—Atherosclerosis
—Gum infections
—Illness such as peptic ulcers, sinusitis, asthma, colds, influenza and pneumonia
—Decreased sense of taste and smell
—Increased headaches and stomach problems
—Chronic bronchitis and emphysema
—Accidental injury due to fires
—Plaque buildup on teeth
—Increased toxicity of other chemical substances such as alcohol, coal dust, cotton fiber, industrial dyes

Health Benefits of Smoking Cessation

—Drop in carbon monoxide levels in the blood
—Heart rate and diastolic blood pressure go down

—Heart and lungs begin to repair damage done by smoking
—Headaches and stomach problems decrease within a few weeks
—People feel more energetic
—Increased sense of taste and smell
—"Smoker's hack" will disappear
—Increased resistance to colds, influenza, and sinusitis
—Within a year, the risk of heart attack will be sharply reduced
—Chronic bronchitis often clears up
—Ulcers are more likely to heal
—Risk of thrombosis will be much less
—Quitting slows the progress of emphysema
—Within 5-10 years after quitting, the risk of getting smoking-related cancers will be considerably reduced
—By 10-15 years after quitting, health risks for an ex-smoker are about the same as are risks for someone who never smoked

Special Concerns for Older Adults

Since many people are aware that most of the ill effects of smoking are cumulative over time, older adults may assume that since they have smoked for many years they will derive no benefit by giving up smoking.

Purpose of Health Promotion

To encourage people to examine the facts about cigarette smoking, to recognize the value of smoking cessation, and provide practical skills for the person who wants to either achieve and/or maintain a nonsmoking lifestyle.

Topics for Health Promotion Classes and Activities

—Information about the risks of cigarette smoking
—Self-identification of smoking patterns and obstacles to smoking cessation
—Motivation to quit smoking
—Personal support for smoking cessation
—Things to do instead of smoking
—Stress reduction and coping skills
—Maintenance skills

Resources for the Person Who Wants to Quit

Contact local chapter for information, films, posters, booklets, or speakers on risks of smoking, especially cancer risk. Works to encourage public policies that support not smoking. Offers stop-smoking classes and printed information on quitting techniques. Trains volunteers to conduct stop-smoking classes. All services are free.

American Cancer Society, Inc.
777 Third Avenue
New York, NY 10017

Contact local chapters for stop smoking programs.

American Heart Association
7320 Greenville Avenue
Dallas, TX 75231

Stop-smoking classes, pamphlets on smoking and health risks to lungs. Contact local chapter for 20-day quit smoking plan and maintenance (1-year) plan for quitters; available for small fee. Local chapters may provide educational services, training for leaders of stop smoking classes.

American Lung Association
1720 Broadway
New York, NY 10019

Information on cancer and pamphlets on quitting smoking. Also has *Smoking and Health: An Annotated Bibliography of Public and Professional Education Materials.* Pub. No. 79-1841, NIH.

Office of Cancer Communications
National Cancer Institute
National Institutes of Health
Bethesda, MD 20014

Contact local churches for stop smoking classes. Little or no fee.

General Headquarters
5-Day Plan to Stop Smoking
Seventh Day Adventist Church
Narcotics Education Division
6840 Eastern Avenue, N.W.
Washington, DC 20012

SmokeEnders and Schick Laboratories both offer stop smoking programs in several cities. Cost of these programs may be more than someone on a fixed income can afford to pay.

WEIGHT CONTROL

Facts Relevant to Older Adults

—80-90% of American adults are at least 5% overweight.
—Approximately 60 million adult Americans are between 5-15 pounds overweight.
—Americans gain an average of 1-2 pounds every year from age 20-50.
—Basic metabolic rate decreases 16% between age 30 and age 70, lessening the amount of calories required to maintain weight.
—The typical American diet is not nutritious.
—The amount of exercise decreases as Americans age.
—Fat increasingly replaces muscle as exercise decreases.

Health Risks for the Overweight

—High blood pressure
—Diabetes
—Cardiovascular disease
—Gall bladder disease
—Shortness of breath
—High blood cholesterol levels
—Osteoarthritis
—Reduced hearing capability
—Breast and uterine cancer
—Sleep disturbances

Special Concerns for Older Adults

The need for information on good nutrition and sound eating habits is compounded by the need to adapt diet and exercise to process of aging.

Purpose of Health Promotion

Teach older adults to adopt eating and exercise habits that will help them reach and maintain desired weight.

Topics for Health Promotion Classes and Activities

—Information on caloric intake and needs
—Information on the nutritional value of foods
—Food monitoring skills
—Retraining eating habits
—Convenience foods
—Supports for change in our everyday environment
—Relationship of exercise and overweight

Resources for Weight Control

Conducts research, offers professional patient and public education programs. Contact local chapters for information and educational services.

American Diabetes Association
2 Park Avenue
New York, NY 10016

Offers a packaged program on nutrition for presentation to groups (specifically older persons) on nutrition. Explains age-related changes in nutritional needs; gives guidelines for reducing food costs. Is designed to be presented in 1 hour; includes instructions, script, a short film, handouts, and provides guidelines for recruiting a local resources person.

National Retired Teachers Association/
American Association of Retired Persons
1909 K. Street N.W.
Washington, DC 20049

Local chapters offer weekly support groups, diet guidelines, education about nutrition and weight control. Dues charged at meetings.

Weight Watchers International, Inc.
800 Community Drive
Manhasset, NY 11030

Check for other local groups like Take Off Pounds Sensibly (TOPS) or Overeaters Anonymous. Many of the major life/health insurance companies have materials on weight control.

References

Wellness in Old Age

FallCreek, S. and Stam. S. B. (Eds.). *The Wallingford Wellness Project—An innovative health promotion program with the elderly.* Seattle, WA: Center for Social Welfare Research, School of Social Work, University of Washington, 1981.

Healthy people: The Surgeon General's report on health promotion and disease prevention. Washington, D.C.: U.S. Department of Health, Education, and Welfare, Public Health Service No. 79-55071, 1979.

Lalonde, B. and FallCreek, S. Preliminary evaluation of the Wallingford Wellness Project: A model health promotion program for the elderly. Seattle, WA: School of Social Work, University of Washington, 1981.

Lowe, J. *Viewpoint: Toward a healthier America.* W. K. Kellogg Foundation, January 1980.

Social and Economic Characteristics of the Older Population. U.S. Department of Commerce Bureau of the Census, 1978.

U.S. Senate. A report to the Special Committee on Aging. *Developments in aging: 1978,* 1979.

Wallingford Wellness Project Staff Team. *Health promotion educational materials.* Seattle, WA: Center for Social Welfare Research, School of Social Work, University of Washington, 1982.

Program Development Guidelines

Fitting Your Population

Benitez, R. Ethnicity, social policy and aging. In R. H. Davis (Ed.) *Aging: Prospects and issues.* Los Angeles, CA: Ethel Percy Andrus Gerontology Center, University of California, 1977.

Block, M. R. Exiled Americans: The plight of Indian aged in the United States. In D. E. Gelfand & A. J. Kutzik (Eds.), *Ethnicity and aging: Theory, research, and policy.* New York: Springer Publishing Company, 1979.

Brody, E. *Long-term care of older people.* New York: Human Sciences Press, 1977.

Chen, P. N. Continuing satisfying life patterns among aging minorities. *Journal of Gerontological Social Work*, 1980, *2*, 199-211.

Cheng, E. *The elder Chinese.* San Diego, CA: The Campanile Press, 1978.

Dukepoo, F. C. *The elder American Indian.* San Diego, CA: The Campanile Press, 1978.

Graham, S. and Reeder, L. G. Social epidemiology of chronic diseases. In H. E. Freeman, S. Leving, & L. G. Reeder (Eds.), *Handbook of medical sociology.* Englewood Cliffs, NJ: Prentice-Hall, 1979.

Ishizuka, K. C. *The elder Japanese.* San Diego, CA: The Campanile Press, 1978.

Jackson, J. J. *Minorities and aging.* Belmont, CA: Wadsworth Publishing Co., 1980.

Jacques, G. Cultural health traditions: A Black perspective. In M. F. Branch & P. P. Paxton (Eds.), *Providing safe nursing care for ethnic people of color.* New York: Appleton Century Crofts, 1976.

Joe, J., Gallerito, C., and Pino, J. Cultural health traditions: American Indian perspectives. In M. F. Branch & P. P. Paxton (Eds.), *Providing safe nursing care for ethnic people of color.* New York: Appleton Century Crofts, 1976.

Kane, R. L. and Kane, R. A. Care of the aged: Old problems in need of new solutions. *Science*, May 1978, *200*(26).

Levkoff, S., Pratt, C., Esperanze, R., and Tomine, S. *Minority elderly: A historical and cultural perspective.* Corvallis, OR: Oregon State University Extension Service, 1979.

Levy, J. E. The older American Indian. In E. G. Youmans (Ed.), *Older Rural Americans: A sociological perspective.* Lexington, KY: University of Kentucky Press, 1967.

Maldonado, D., Jr. The Chicano aged. *Social Work*, 1975, *20*, 213-216.

Moore, J. W. Mexican-Americans. *The Gerontologist*, 1971, *2*, 30-35.

National Institute on Aging, "Minorities and How They Grow Old," *Age Page*, Bethesda, MD 20205, August, 1980.

Palmore, E. Total chance of institutionalization among the aged. *The Gerontologist*, 1976, *16*, 504-507.

Rodriguez, D. P. and Quintero, J. H. Cultural health traditions: The Latino/Chicano perspective. In M. F. Branch & P. P. Paxton (Eds.), *Providing safe nursing care for ethnic people of color.* New York: Appleton Century Crofts, 1976.

Rudov, Melvin, H. and Santangelo, Nancy. *Health status of minorities and low income groups.* U.S. Department of Health, Education, and Welfare, Public Health Services, Health Resources Administration, 1979.

U.S. Bureau of the Census, *Current Population Reports*, Special Studies Series, 1974.

Valle, R. and Mendoza, L. *The elder Latino.* San Diego, CA: The Campanile Press, 1978.

Wagner, C. J. and Rabeau, E. S. *Indian poverty and Indian health.* Health, Education, and Welfare Indicators, March 1964, 24-44.

Watson, W. H. Older Blacks in the rural South. *Minority Aging Exchange Newsletter*, University Center on Aging, San Diego State University, March, 1979.

The Pillars of Health Promotion

Harris, L. and Associates. *Health Maintenance.* Pacific Mutual Life Insurance Co., 1978.

Healthy people: The Surgeon General's report on health promotion and disease prevention. Washington, D.C.: U.S. Department of Health, Education, and Welfare, Public Health Service No. 79-55071, 1979.

Keelor, R. Pep up your life: A fitness book for seniors. Hartford, CT: The Travelers Insurance Companies, 1980.

Variations on a Basic Theme

Self-Care

Butler, R., et al. Self-care, self-help, and the elderly. *International Journal of Aging and Human Development*, 1979-1980, *10*(1), 96-97.

Farquhar, J. W., et al. Community education for cardiovascular health. *Lancet*, 1977, June 4, 1192-1195.

Levin, L. S. The layperson as primary health care practitioner. *Public Health Reports*, 1976, *91*(3), 206.

Levin, L. S. Patient education and self-care: How do they differ? *Nursing Outlook*, 1978, *26*, 170-175.

Levin, L. S., Katz, A. H., and Holst, E. *Self-care: Lay initiatives in health* (2nd ed.). New York: Prodist, 1979.

Norris, C. Self-care. *American Journal of Nursing*, 1979, *79*(3), 486.

Sehnert, K. W., Awkward, B., and Lesage, D. *The senior citizen as an "activated patient": A course guide.* HEW-100-75-0116. Center for Continuing Health Education, School for Summer and Continuing Education, Georgetown University: September 1976.

Sehnert, K. W. and Eisenberg, H. *How to be your own doctor (sometimes).* New York: Grosset & Dunlap, 1975.

U.S. Department of Health and Human Services, Public Health Services. *Focal points.* Atlanta, GA: Centers for Disease Control, Bureau of Health Education, 1980.

Williamson, J.D. and Danaher, K. *Self-care in health.* London, England: Croom Helm, 1978.

Peer Advocacy

Bolton, C. R. and Dignum-Scott, J. E. Peer-group advocacy counseling for the elderly: A conceptual model. *Journal of Gerontological Social Work*, 1979, *1*(4), 322.

Cox, C. A pilot study: Using the elderly as community health educators. *International Journal of Health Education*, 1979, *22*(1), 49-52.

Ivey, A. E. *Microcounseling: Innovations in Interview Training.* Springfield, IL: Charles C. Thomas, 1971.

Support Networks

Biegel, D. E. and Sherman, W. R. Neighborhood capacity building and the ethnic aged. In D. E. Gelfand & A. J. Kutzik (Eds.), *Ethnicity and aging.* New York: Springer Co., 1979.

Brody, S. J., Paulschok, S. W., & Maciocchi, C. F. The family caring unit: A major consideration in the long-term support system. *The Gerontologist*, 1978, *18*(6), 556-561.

Collins, A. H. and Pancoast, D. L. *Natural helping networks: A strategy for prevention.* Washington, D.C.: National Association of Social Workers, 1976.

Ehrlich, P. *Mutual help for community elderly demonstration and research project: The mutual help model* (Vol. I). Carbondale, IL: Southern Illinois University, 1979.

Ehrlich, P. *Mutual help for community elderly: The mutual help model* (Vol. II). Carbondale, IL: Southern Illinois University, 1979.

Ehrlich, P. Service delivery for the community elderly: The mutual help model. *Journal of Gerontological Social Work*, 1979, *2*(2), 125-135.

Evans, L. K. Maintaining social interaction as health promotion in the elderly. *Journal of Gerontological Nursing*, 1979, *5*(2), 19-21.

Hess, B. B. Self-help among the aged. *Social Policy*, 1976, *7*, 55.

Lopata, H. Z. Support systems of elderly urbanites: Chicago of the 1970's. *The Gerontologist,* 1975, *15*(1), (Part 1), 35.

Mannery, J. D. Aging in American society. In G. Sorenson & J. Tift, *Meeting the psychosocial needs of the older person.* Minneapolis, MN: The Program on Aging of Augsburg College, 1977.

Monk, A. Family supports in old age. *Social Work*, 1979, *24*(6), 534.

Shanas, E. The family as a social support system in old age. *The Gerontologist*, 1979, *19*(2), 169.

Stephens, R. C., et al. Aging, social support systems and social policy. *Journal of Gerontological Social Work*, 1978, *1*(1), 40.

Toseland, R. W., et al. A community outreach program for socially isolated older persons. *Journal of Gerontological Social Work*, 1979, *1*(3), 217-219.

Treas, J. Family support systems for the aged: Some social and demographic considerations. *The Gerontologist*, 1977, *17*(6), 487-488.

Community Activation

Miller, I. and Solomon, R. The development of group services for the elderly. *Journal of Gerontological Social Work*, 1980, *2*(3), 247-248.

Tips for Facilitators

Beal, G. M., Bohlen, J. M., and Raudabaugh, J. N. *Leadership and dynamic group action.* Ames, IA: Iowa State University Press, 1962.

Bloom, B. S. (Ed.). *Taxonomy of educational objectives, handbook I: Cognitive domain.* New York: David McKay Company, Inc., 1956.

Burnside, I. M. Group work with the aged: Selected literature. *The Gerontologist*, 1970, *10*(3), 241-246.
Curry, R. C. *Training for trainers: Serving the elderly the technique, part 6.* Durham, NH: New England Gerontology Center, 1980.
Davis, L. N. *Planning, conducting, and evaluating workshops.* Austin, TX: Learning Concepts, Inc., 1974.
Olmstead, J. A. *Small group instruction: Theory and practice.* Alexandria, VA: Human Resources Research Organization, 1974, 81.
Petty, B. J., et al. Support groups for elderly persons in the community. *The Gerontologist*, 1976, *16*(6), 522-528.
Worsley, A. F. Improving classroom discussion: Ten principles. *Improving College and University Teaching*, 1975, *23*, 27.

Class Outlines

Stress Management

Ardell, D. *14 days to a wellness lifestyle.* Mill Valley, CA: Whatever Publications, Inc., 1982.
Farquhar, J. *The American way of life need not be hazardous to your health.* New York: W. W. Norton, 1978.
Girdano, D. and Everly, G. *Controlling stress and tension.* Englewood Cliffs, NJ: Prentice-Hall, 1979.
Jacobson, E. *You must relax.* New York: McGraw-Hill, 1978.
McCamy, J. and Presley, J. *Human lifestyling.* New York: Harper and Row, 1975.
Selye, H. *Stress without distress.* New York: J. B. Lippincott, 1974.

Nutrition

Farquhar, J. *The American way of life need not be hazardous to your health.* New York: W. W. Norton, 1978.
Gutherie, H. A. *Introductory nutrition.* St. Louis, MO: C. V. Mosby Co., 1971.
Williams, J. and Silverman, G. *No salt, no sugar, no fat.* Concord, CA: Nitty Gritty Productions, 1981.
U.S. Department of Agriculture, U.S. Department of Health, Education and Welfare. *Nutrition and your health: Dietary guidelines for Americans*, 1981.
U.S. Senate. Select Committee on Nutrition and Human Needs. *Dietary goals for the United States*, 2nd edition, 1977.

Physical Fitness

Ardell, D. *14 days to a wellness lifestyle.* Mill Valley, CA: Whatever Publications, Inc., 1982.
Harris, L. and Associates. *Health maintenance.* Pacific Mutual Life Insurance Co., 1978.
Harris, R. *Guide to fitness after fifty.* New York: Plenum, 1977.
Keelor, R. Pep up your life: A fitness book for seniors. Hartford, CT: The Travelers Insurance Companies, 1980.

Personal and Community Self-Help

Bloom, L. Z., Coburn, K., and Pearlman, J. *The new assertive woman.* New York: Dell Publishing Co., Inc., 1975.
Gambrill, E. D. and Richey, C. A. An assertion inventory for use in assessment and research. *Behavior Thearpy*, 1975, *6*, 550-561.
Richey, C. A. *Workshop manual for trainers.* Seattle, WA: University of Washington, 1979.
Smith, M. J. *When I say no, I feel guilty.* New York: Bantam Book and the Dial Press, 1975.

Common Health Concerns

Accidental Injuries

Accidents and the elderly. *Age Page*, National Institute on Aging, July 1980.
A winter hazard for the old: Accidental hypothermia. National Institute on Aging, 1980.
Heat, cold, and being old. *Age Page*, National Institute on Aging, February 1981.
Older consumers and stairway accidents. Fact Sheet No. 48, U.S. Consumer Product Safety Commission, 1975.
Price, J. H. Unintentional injury among the aged. *Journal of Gerontological Nursing*, 1978, *4*, 36-40.

Alcoholism

Alcohol and the elderly. Fact Sheet, National Institute of Alcohol Abuse and Alcoholism, 1978.
Rathbone-McCuan, E. and Triegaardt, J. The older alcoholic and the family. *Alcohol Health and Research World*, Summer 1979, 7-12.
Schuckit, M. A. Sensitivity complicates elders' substance abuse. *Generations*, 1980, *5*(2), 8-9;36.
Schuckit, M. A. Geriatric alcoholism and drug abuse. *Gerontologist*, 1977, *17*(2), 168-174.
Schuckit, M. A. and Pastor, P. The elderly as a unique population: Alcoholism. *Alcoholism: Clinical and Experimental Research*, 1978, *2*(1), 31-38.

Arthritis

Arthritis, the basic facts. Anon., The Arthritis Foundation, 1978.
Freese, A. A. *Arthritis: Everybody's disease.* Public Affairs Pamphlet #562.
Lorig, K. and Fries. *The arthritis helpbook.* Addison-Wesley, 1980.
Skeist, R. J. *To your good health.* Chicago Review Press, 1980. (Chapter 5).

Cancer

Peterson, B. H., Kennedy, B. J., Butler, R. N., and Gastel, B. *Aging and cancer management.* New York: American Cancer Society, 1979.
Understanding the cancer patient. New York: American Cancer Society, 1980.
Wynder, E. (Ed.). *The book of health.* New York: Franklin Watts, 1981.

Diabetes

Biermann, J. and Toohey, B. *The diabetics total health book.* Los Angeles: J. P. Tarcher, Inc., 1980.
Duncan, T. G. Teaching commonsense health care habits to diabetic patients. *Geriatrics*, Oct. 1976, *31*(10), 93-96.
Glickman, S. and Lipshutz, J. *Your health and aging.* New York: Division of Gerontology, Office of Urban Health Affairs, NYU Medical Center, 1981, 99-105.
Helfand, Arthur E., D.P.M. Common foot complications in the elderly diabetic. *Journal of the American Podiatry Association*, June 1977, *67*(6), 406-408.
Moss, J. M. Pitfalls to avoid in diagnosing diabetes in elderly patients. *Geriatrics*, Oct. 1976, *31*(10), 52-55.
Skeist, R. *To your good health.* Chicago: Chicago Review Press, 1980, 121-126.
Thomas, K. P., M.Ed. Diabetes Mellitus in elderly persons, *Nursing Clinics of North America*, March 1976, *11*(1), 157-167.

Foot Care

Patient instructions for the care of the diabetic foot. San Francisco, CA: Handout, California College of Podiatric Medicine.

Rossman, I. (Ed.). *Clinical geriatrics* (2nd ed.). Philadelphia, PA: J. B. Lippincott, Co., 1979, 638-639.
Skeist, R. (Ed.). *To your good health.* Chicago Review Press, 1980, 103-109.
Ventura, E. Foot care for diabetics. *American Journal of Nursing*, 1978, *75*, 886-888.

Heart Condition

Farquhar. J. W. *The American way of life need not be hazardous to your health.* New York: W. W. Norton & Co., 1978.
Guyton, A. C. *Basic human physiology: Normal function and the mechanisms of diesease.* Philadelphia, PA: W. B. Saunders Co., 1977.
Heart to heart. New York: Pamphlet, Life Extension Institute.
Rossman, I. (Ed.). *Clinical geriatrics* (2nd ed.). Philadelphia, PA: J. B. Lippincott, 1979.
Why risk heart attack? Pamphlet, American Heart Association.

Hypertension

Giblin, E. Controlling high blood pressure. *American Journal of Nursing*, 1978, *75*, 824.
Ignacio, A., et al. Multidisciplinary team approach to patient education in hypertension. *Hawaii Medical Journal*, 1976, *35*, 298-300.
Long, M., et al. Hypertension: What patients need to know. *American Journal of Nursing*, 1976, *73*, 765-767.
Statement on hypertension in the elderly. National High Blood Pressure Information Center, National Institutes of Health, Bethesda, 1980. In I. Rossman (Ed.), *Clinical geriatrics*, (2nd ed.). Philadelphia, PA: J. B. Lippincott Company, 1979.
Taylor, D. W., et al. Compliance with antihypertensive drug therapy. *Annals*, New York Academy of Sciences, 1978, 390-403.
Ward, G., et al. Treating and counseling the hypertensive patient. *American Journal of Nursing*, 1978, *76*, 824-828.

Insomnia

Castleman, M. How to get a good night's sleep. *Medical Self-Care*, Winter 1981 (15), 22-26.
Hales, D. *The complete book of sleep.* Reading, MA: Addison-Wesley, 1981.
Secrets of sleep. *Newsweek*, July 13, 1981, *98*, 48-53.
Skeist, R. J. *To your good health!* Chicago: Chicago Review Press, 1980, 57-64.

Medicine Management

Graedon, J. What you need to know about drugs. *Medical Self-Care*, 1979-80, *7*, 30-36.
Lambing, M. L., et al. *Idea exchange: Preventing misuse of drugs by elders.* Paper presented at Western Gerontological Society Annual Meeting, Anaheim, California, March 1980.
Long, J. W. *The essential guide to prescription drugs.* New York: Harper and Row, 1977.
Safe use of medicine by older people. *Age Page*, National Institute on Aging, November 1980.
Vestal, R. E. *Drugs and the elderly. National Institute on Aging*, Science Writer Seminar Series, Pub. #79-1449, July 1979.

Normal Changes of Aging

Hendricks, J. and Hendricks, C. D. *Aging in mass society: Myths and realities.* Cambridge, MA: Winthrop Publishers, Inc., 1977.
Schrock, M. *Holistic assessment of the healthy aged.* New York: John Wiley and Sons, 1980.

Oral Health

Chauncey, H. H. and House, J. Dental problems in the elderly. *Hospital Practice*, December 1977, 81-86.

Rossman, I. (Ed.). *Clinical geriatrics* (2nd ed.). Philadelphia, PA: J. B. Lippincott, Co., 1979, 618-619.

Ship, I. I. Geriatric dentistry must meet elder's changing needs. *Generations*, 1980, *5*, 27-37.

Sensory Loss

The aging eye: Facts on eye care for older persons. Anon., pamphlet, National Society to Prevent Blindness, 1979.

Ernst, M. Sensory impairment increases with age: Societal attitudes hinder compensation. *Generations*, 1980, *5*.

Freese, A. S. *Glaucoma—diagnosis, treatment, prevention.* Public Affairs Pamphlet #568, New York.

Freese, A. S. *Cataracts and their treatment*. Public Affairs Pamphlet #545, New York.

Greyton, A. C. *Human physiology: Normal function and mechanisms of disease.* Philadelphia, PA: W. B. Saunders Company, 1977.

Hearing loss in the aged. Anon., Environmental Communication Intervention for the Aging Project, Murray State University, Kentucky.

Shore, H. Designing a training program for understanding sensory losses in aging. *The Gerontologist*, 1976, *16*.

Tobacco Smoking

Dangers of smoking—benefits of quitting. American Cancer Society, New York, revised, 1980.

The smoking digest: Progress report on a nation kicking the habit. U.S. Department of Health, Education, and Welfare; Public Health Service; National Institute of Health Publication No. 79-1549, 1979. Available from the Office of Cancer Communications, National Cancer Institute.

Weight Control

Busse, E. W., M.D. Eating in late life: Physiologic and psychologic factors. *New York State Journal of Medicine*, August 1980, 1496-1497.

Diabesity—A new name for an old disease. *Taking Care*, Newsletter of The Center for Consumer Health Education, May 1980, *2*(5), 1-2.

Farquhar, J. *The American way of life need not be hazardous to your health.* New York: W. W. Norton & Co., 1978.

Healthy people: The Surgeon General's report on health promotion and disease prevention. Washington, D.C.: U.S. Department of Health, Education, and Welfare, Public Health Service, No. 79-55071, 1979, 129-134.

Sebrell, W. H. A rational approach to weight control. Pamphlet from Weight Watchers International Inc., 1978.

Successful diet and exercise therapy conducted in Vermont for diabesity. *Journal of the American Medical Association*, 1980, *243*.

Index